Augmented Intelligence Toward Smart Vehicular Applications

Augmented Intelligence Toward Smart Vehicular Applications

Edited by
Nishu Gupta, Joel J. P. C. Rodrigues, and
Justin Dauwels

CRC Press
Taylor & Francis Group
Boca Raton London New York

CRC Press is an imprint of the
Taylor & Francis Group, an **informa** business

First edition published 2021 by CRC Press
6000 Broken Sound Parkway NW, Suite 300, Boca Raton, FL 33487-2742

and by CRC Press
2 Park Square, Milton Park, Abingdon, Oxon, OX14 4RN

© 2021 Taylor & Francis Group, LLC

CRC Press is an imprint of Taylor & Francis Group, LLC

Library of Congress Cataloging-in-Publication Data

Names: Gupta, Nishu, editor. | Rodrigues, Joel José P. C., editor. | Dauwels, Justin, 1977- editor.
Title: Augmented intelligence toward smart vehicular applications / edited by Nishu Gupta, Joel J. P. C. Rodrigues & Justin Dauwels.
Description: First edition. | Boca Raton : CRC Press, 2021. | Includes bibliographical references and index.
Identifiers: LCCN 2020028979 (print) | LCCN 2020028980 (ebook) | ISBN 9780367435462 (hardback) | ISBN 9781003006817 (ebook)
Subjects: LCSH: Intelligent transportation systems. | Automated vehicles.
Classification: LCC TE228.3 .A92 2021 (print) | LCC TE228.3 (ebook) | DDC 629.04/60285--dc23
LC record available at https://lccn.loc.gov/2020028979
LC ebook record available at https://lccn.loc.gov/2020028980

ISBN: 9780367435462 (hbk)
ISBN: 9781003006817 (ebk)

Typeset in Palatino
by Deanta Global Publishing Services, Chennai, India

Dr. Nishu Gupta dedicates this book to his

teachers, family, and friends

Prof. Joel J. P. C. Rodrigues dedicates this book to his

wife Charlene Rodrigues and children Catarina, José, and Carolina

If I have seen further it is by standing on the shoulders of Giants.

– Isaac Newton

Prof. Justin Dauwels dedicates this book to his

wife Shoko Dauwels and children Lien and Louis

Contents

Part 1   Introduction

Part 2   Designing and Evaluation

Part 3 Smart Safety Measures and Applications

Foreword

by Dr. Xabiel García Pañeda, PhD

Associate Professor, Department of Informatics, University of Oviedo, Gijón-Xixón, Asturies, Spain

I am pleased to write this Foreword because I feel *Augmented Intelligence Toward Smart Vehicular Applications* deeply emphasizes the state-of-the art technologies that comprise many research explorations in the field of artificial intelligence and that it can offer various vehicular applications through networking and automation.

Although the focus of this work is on augmented intelligence, it contains much more that will be of interest to those outside this subject. The range of topics covered in this book is quite extensive and each one is discussed by experts in their own fields. The advances and challenges are discussed with a focus on successes, failures and lessons learned, open issues, unmet challenges and future directions. In this book, I see that the editors and authors have conceptualized the intellectual foundations of intelligent transportation, elaborated its distinctive pedagogy in vehicular technology applications, and studied its patterns and impact on the common man. It will certainly help researchers and professional practitioners develop a shared vision and understanding of interpretive discussion.

We encounter the challenges faced in addressing technological problems such as providing hassle-free, safe and enjoyable driving experiences. These challenges are both difficult and interesting. Researchers are working on them to develop new approaches and provide new solutions to keep up with the ever-changing potential threats. This book is a good initiative in that direction. I believe it will provide a solid platform to various realms of the existing and upcoming technologies, especially in the fields of artificial intelligence, human-machine interaction, autonomous vehicles and intelligent transportation. The authors can be confident that there will be many grateful readers who will have gained a broader perspective of the disciplines of augmented intelligence and its applications because of their efforts. I hope the book will serve as a primer for industry and academia, professional developers and upcoming researchers across the globe to learn, innovate and realize the multifold capabilities of AI and vehicular applications.

I wish good luck to the editors and contributors of the book.

Foreword

by Dr. Neeraj Chaudhary

Technical Officer, CSIR-National Physical Laboratory (India), Government of India

I am delighted to write the Foreword for *Augmented Intelligence Toward Smart Vehicular Applications*, which highlights the importance that artificial intelligence technology can offer in taking care of various vehicular applications through networking and automation. It intends to demonstrate to its readers useful vehicular applications and architectures that cater to their improved driving requirements. The application domains have a large range in which vehicular networking, communication technology, sensor devices, computing materials and devices, Internet of Things (IoT) communication, intelligent transportation, vehicular and on-road safety, data security, and other trendy topics are included.

Augmented intelligence is a trending junction where IoT meets vehicular technology. The idea of connected and autonomous vehicles continues to be the intersection of automotive intelligence and IoT innovation. The automotive industry continues to evolve and enable the era of intelligent mobility as the autonomous vehicles space makes impressive strides and new technology emerges before us.

This book provides a window to the research and development in the field of vehicular communication in a comprehensive way and enumerates the evolutions of contributing tools and techniques. The range of topics covered in this book is quite extensive, and every topic is discussed by experts in their respective fields. I am confident that this book will provide an effective learning experience and a practical reference for researchers, professionals, and students who are interested in the integration of artificial intelligence in vehicular applications and its advances in the engineering field.

The authors can be assured that because of their contributions there will be many readers who will gain a better understanding of artificial intelligence and its applications in the automobile industry. I highly recommend this book to a variety of audiences including: academicians; industrial engineers; researchers in the field who use transportation systems, vehicular communication, and IoT; communication technology specialists; as well as ad hoc communication network students and scholars.

Preface

Augmented intelligence in vehicular networks is the convergence of artificial intelligence (AI) and multimedia. AI has already been unanimously accepted as a revolutionary solution to future technological challenges. By automating and computerizing vehicles, we can make our mobility more secure, quicker, less expensive, cleaner, and progressively pleasant.

Written by international experts, this book intends to present to its readers smart vehicular applications and paradigms by augmenting intelligence and their subsequent prototypes. This book discusses various dimensions of vehicular communication in intelligent transportation systems (ITS) and explores its trendy approaches and diverse technologies. It focuses on the organization of AI and aspects of deep learning algorithms, particularly multimodal transport. New approaches to the prediction of road safety needs are also considered. The chapters of the book provide insight into the importance that machine learning solutions can have in taking care of vehicular safety through internetworking and automation. Key features of this book are the inclusion and elaboration of recent and emerging developments in various specializations of ITS.

The book is divided into different parts, each having multiple chapters. The contents of the book have been organized in a reader-friendly manner. It caters to a wide audience including university professors, graduates and PhD students, industry professionals, and researchers, particularly in the field of vehicular communication and sensor network. The contents are presented in three main parts: Introduction, designing and evaluation, and smart safety measures and applications. The first part presents an overview of augmented intelligence, edge computing, and vehicular ad hoc networks (VANETs). The second part focuses on designing, management, and evaluation of various vehicular models; simulation for autonomous vehicles; application testing; and approximation algorithm toward smart vehicular applications. Finally, the third part highlights Internet of things (IoT) solutions in vehicular safety, data structures, and mobility techniques for ITS.

This book attracted contributions from all over the world, and we would like to thank all the authors for submitting their work. We extend our appreciation to the reviewers for their timely and focused review comments. We gratefully acknowledge all the authors and publishers of the books quoted in the references.

Nishu Gupta
Vaagdevi College of Engineering, Warangal, India

Joel J. P. C. Rodrigues
Federal University of Piauí, Brazil; Instituto de Telecomunicações, Portugal

Justin Dauwels
Nanyang Technological University, Singapore

MATLAB® is a registered trademark of The MathWorks, Inc. For product information, please contact:

The MathWorks, Inc.
3 Apple Hill Drive
Natick, MA 01760-2098 USA
Tel: 508 647 7000
Fax: 508-647-7001
E-mail: info@mathworks.com
Web: www.mathworks.com

Editors

Nishu Gupta is a senior member of the Institute of Electrical and Electronics Engineers (IEEE). He specializes in the field of vehicular communication and networking. He is a postdoctoral fellow at the University of Oviedo, Spain, under the research group on systems for multimedia and the Internet of Things (SMIOT). Currently, he is associate professor and serving as the head of the Electronics and Communication Engineering Department at Vaagdevi College of Engineering, Warangal, India. He earned his PhD from Motilal Nehru National Institute of Technology Allahabad, Prayagraj, India, which is an institute of national importance as declared by the Government of India. Dr. Gupta works in collaboration with Nokia Pvt. Ltd. as an Invention Partner. He has been designated as Conference Chair at the 2020 EAI/Springer International Conference on Cognitive Computing and Cyber Physical Systems (IC4S 2020). He has supervised several final projects of bachelor and master's theses in his main line of work and has published three patents, many in process. He has authored and edited several books with international publishers, such as Taylor & Francis, Springer, Wiley, Scrivener, and Cambridge Scholar Publishing. Dr. Gupta is on the editorial board of various reputable journals and transactions. He serves as an active reviewer for various journals like *IEEE Transactions on Intelligent Transportation Systems, IEEE Access, IET Communications*, and many more. At present, he is working in collaboration with multiple academicians and researchers across the globe. He is the recipient of the Best Paper Presentation Award presented at an international conference at Nanyang Technological University, Singapore. He serves as Chief Coordinator, Institute of Innovation Cell, under MHRD-IIC, Government of India, in addition to holding many other key positions in the academic and research fields. His areas of research interests include: autonomous vehicles, edge computing, augmented intelligence, IoT, Internet of vehicles, deep learning, ad-hoc networks, vehicular communication, driving efficiency, cognitive computing, human–machine interaction, and traffic pattern prediction.

Joel J. P. C. Rodrigues is a professor at the Federal University of Piauí, Brazil; senior researcher at the Instituto de Telecomunicações, Portugal; and collaborator of the postgraduate program on teleinformatics engineering at the Federal University of Ceará (UFC), Brazil. Prof. Rodrigues is the leader of the Next Generation Networks and Applications (NetGNA) research group (CNPq), an IEEE Distinguished Lecturer, Member Representative of the IEEE Communications Society on the IEEE Biometrics Council, and the president of the scientific council at ParkUrbis – Covilhã Science and Technology Park. He was Director for Conference Development at the IEEE

ComSoc Board of Governors, Technical Activities Committee Chair of the IEEE ComSoc Latin America Region Board, a Past-Chair of the IEEE ComSoc Technical Committee on eHealth, a Past-Chair of the IEEE ComSoc Technical Committee on Communications Software, a Steering Committee Member of the IEEE Life Sciences Technical Community, and Publications Co-Chair. He is the Editor-in-Chief of the *International Journal on E-Health and Medical Communications* and Editorial Board Member of several highly-reputed journals. He has been the General Chair and Technical Program Committee Chair of many international conferences, including the IEEE International Conference on Communications; IEEE Global Communications Conference; IEEE International Conference on E-health Networking, Application and Services; and IEEE Latin-American Conference on Communications. He has authored or co-authored over 900 papers in refereed international journals and conferences, three books, two patents, and one ITU-T recommendation. He has been awarded several outstanding leadership and outstanding service awards by the IEEE Communications Society and several best papers awards. Prof. Rodrigues is a member of the Internet Society, a senior member of the Association for Computing Machinery, and fellow of the IEEE.

Justin Dauwels is an associate professor of the School of Electrical and Electronic Engineering at the Nanyang Technological University (NTU) in Singapore. He also serves as deputy director of the ST Engineering–NTU corporate lab, which comprises 100+ PhD students, research staff, and engineers, developing novel autonomous systems for airport operations and transportation. His research interests are in data analytics with applications to intelligent transportation systems, autonomous systems, and analysis of human behavior and physiology. He earned his PhD in electrical engineering at the Swiss Polytechnical Institute of Technology (ETH) in Zurich in 2005. Moreover, he was a postdoctoral fellow at the RIKEN Brain Science Institute (2006–2007) and a research scientist at the Massachusetts Institute of Technology (2008–2010). He has been a Japan Society for the Promotion of Science (JSPS) postdoctoral fellow (2007), a Belgian American Educational Foundation (BAEF) fellow (2008), a Henri-Benedictus fellow of the King Baudouin Foundation (2008), and a JSPS invited fellow (2010, 2011). His research on intelligent transportation systems has been featured by the BBC, *Straits Times*, Lianhe Zaobao, Channel 5, and numerous technology websites. Besides his academic efforts, Dr. Dauwels also collaborates with local start-ups, SMEs, and agencies, in addition to MNCs, in the field of data-driven transportation, logistics, and medical data analytics.

List of Contributors

Anuj Abraham
School of Electrical and Electronic
  Engineering
Nanyang Technological University
  Singapore

B. Amutha
Department of Computer Science
  and Engineering
SRM Institute of Science and
  Technology
Kattankulathur, Chennai, India

M. Baskar
Department of Computer Science
  and Engineering
SRM Institute of Science and
  Technology
Kattankulathur, Chennai, India

Priyanka Bhardwaj
Department of Electronics and
  Communication Engineering
Bharati Vidyapeeth's College of
  Engineering
New Delhi, India

M. Dhana Lakshmi Bhavani
Department of Electronics and
  Communication Engineering
National Institute of Technology
  Silchar
Assam, India

Anirban Bhowmick
Department of Computer
  Application
C.R.T.I. Tinkonia
Goodshed Road, Burdwan, India

Apratim Choudhury
Siemens Mobility
Singapore

Justin Dauwels
School of Electrical and Electronic
  Engineering
Nanyang Technological University
Singapore

Joydeep Dey
Department of Computer Science
M.U.C. Women's College
B.C. Road, Burdwan, India

K. Veena Divya
Department of Electronics &
  Instrumentation Engineering
R.V. College of Engineering
Bengaluru, Karnataka, India

Banishree Ghosh
Agency for Science, Technology and
  Research (A*STAR)
Singapore

Ramkumar Jayaraman
Department of Computer Science
  and Engineering
SRM Institute of Science and
  Technology
Kattankulathur, Chennai, India

Sunil Karforma
Department of Computer Science
The University of Burdwan
Burdwan, India

Hong Li
NXP Semiconductors Eindhoven
Netherlands

Tomasz Maszczyk
Institute of High Performance
    Computing
A*STAR, Singapore

Chetan B. Math
Institute of Infocomm Research
A*STAR, Singapore

Priyanka Mehta
School of Electrical and Electronic
    Engineering
Nanyang Technological University
    Singapore

Usman Muhammad
School of Electrical and Electronic
    Engineering
Nanyang Technological University
    Singapore

R. Murugan
Department of Electronics and
    Communication Engineering
National Institute of Technology
    Silchar
Assam, India

Raghavendra Pal
Department of Electronics
    and Communication
    Engineering
Vaagdevi College of Engineering
Warangal, India

P.M. Rajasree
Department of Electronics &
    Instrumentation Engineering
R.V. College of Engineering
Bengaluru, Karnataka, India

Shyam Sundar Rampalli
School of Electrical and Electronic
    Engineering
Nanyang Technological University
    Singapore

Arindam Sarkar
Department of Computer Science &
    Electronics
R.K.M. Vidyamandira
Belur Math, Belur, India

Shashwat
Energy Research Institute
Nanyang Technological University
Singapore

Surjeet
Department of Electronics and
    Communication Engineering
Bharati Vidyapeeth's College of
    Engineering
New Delhi, India

Nagacharan Teja Tangirala
School of Electrical and Electronic
    Engineering
Nanyang Technological University
    Singapore

M.J. Vidya
Department of Electronics &
    Instrumentation Engineering
R.V. College of Engineering
Bengaluru, Karnataka, India

Pranjal Vyas
School of Electrical and Electronic
    Engineering
Nanyang Technological University
    Singapore

Part 1

Introduction

1

Augmented Intelligence and Edge Computing: Introduction and Trending Features

Priyanka Bhardwaj and Surjeet

CONTENTS

ABSTRACT   Augmented Intelligence (AuI) in vehicular networks is the convergence of artificial intelligence (AI) and multimedia. AI has already been unanimously accepted as a revolutionary solution to future technological challenges. By automating and computerizing vehicles, we can make our mobility more secure, quicker, less expensive, cleaner, and progressively pleasant. This chapter presents a brief introduction to AuI and its relationship with AI. The chapter presents a detailed comparison between AuI and AI and discusses the latest trends, applications, benefits, and future scope of this technology.

1.1 Introduction to Augmented Intelligence

Machines are performing better and better at analyzing data. Due to recent breakthroughs in technology, organizations and businesses are increasingly under pressure to redesign their workflow processes and decide which tasks to automate, which tasks to augment, and which tasks to leave to humans. There are many jobs that machines cannot do, so as machines take over highly repeatable work, people will migrate to roles that call for critical thinking, creativity, judgment, and common sense. As shown in Figure 1.1, augmented intelligence (AuI) unites the strengths of humans and machines [1].

AuI is one of the most important emerging technologies. It covers the subfields of artificial intelligence (AI), machine learning, and neural networks, and it is going to harness the way in which companies and people work. Augmented intelligence, also known as intelligence augmentation or cognitive augmentation, is the next level in AI. The word "augmented" means "to improve" [2].

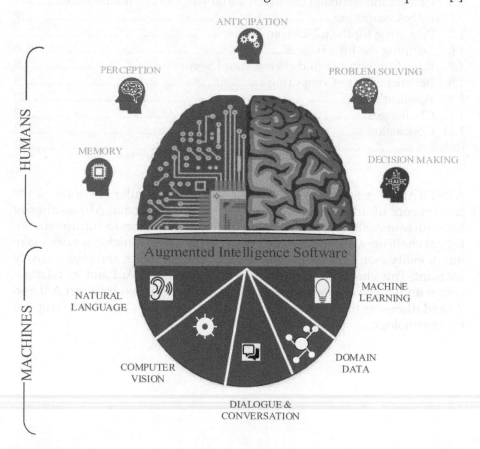

FIGURE 1.1
Augmented intelligence combines human and machine intelligences.

AI software will simply improve products and services and not replace the humans who use them. AuI refers to the creation of a close-to-human autonomous intelligence using modern technology. It describes how normal human intelligence is supplemented through the use of technology. One may say that:

Augmented Intelligence = Human Intelligence + Computer based Intelligence

One may think of AuI as augmented reality (AR), a technology which combines real-world environments with computer-generated information such as images, text, videos, animations, and sound. It has the ability to record and analyze the environment in real time. It is becoming more attractive as a mainstream technology mainly due to the proliferation of modern mobile computing devices like smartphones and tablet computers with location-based services [3]. Like AR, AuI adds layers of information on top of human intelligence, helping humans to function at their best.

The pioneers of augmentation are the sectors that generate a lot of data, such as law, health care, and agriculture. The main objective of AuI is to create an entirely new process automated and designed for 20% manual exceptions. As shown in Figure 1.2, AuI repeats a cycle of understanding, interpretation, reasoning, learning, and assuming [4].

1.2 Artificial Intelligence and Its Applications

Artificial intelligence, as its name implies, is a different form of intelligence to our own. It is human-like intelligence that works in a similar way to our brains [5]. AI is a science with research activities in the areas of image processing, expert systems, natural language processing, and computer vision. Some of these applications are illustrated in Figure 1.3 [6]. AI has been revolutionary and is everywhere: From wearable products to driverless cars. Even the system that prevents a car from starting because a door is open is AI in action. AI, in the forms of machine learning, voice recognition, and predictive analysis, enables robots to provide financial advice [7]. For example, IBM has invested heavily in AI with the Watson cognitive system. Some developed and fast-developing nations like the USA, China, Russia, UK, and France have realized the great potential of AI and are competing to win the AI race.

1.3 Comparing Artificial Intelligence and Augmented Intelligence

AI and AuI are but some of the technologies that are affecting organizations, economies, and our societies. Although AuI and AI are closely related, they

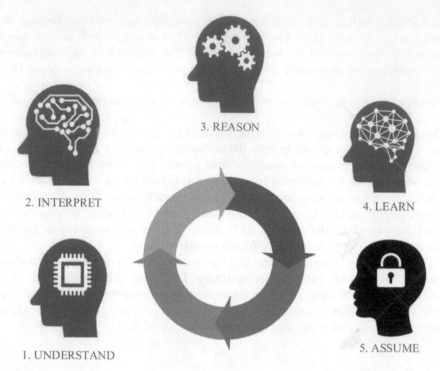

FIGURE 1.2
How augmented intelligence works.

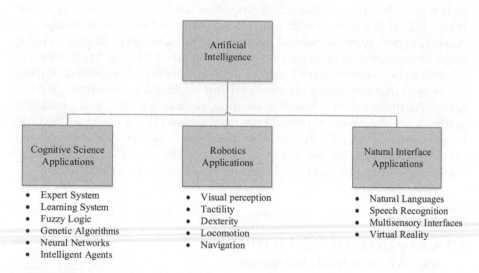

FIGURE 1.3
Some applications of artificial intelligence.

are different. AuI uses AI techniques but keeps the human in the loop. The goal of AuI is not to replace human activities but instead to elevate existing human capabilities. AuI often takes over small repetitive tasks. Focusing just on AI can expose companies to unforeseeable challenges in the future. That is why we must distinguish between AI and AuI. The key difference between AI and AuI lies in the cognitive learning and application of any machine. While AI is meant to help humans with things like computation, memory, perseverance, precision, recognition, and speed, AuI allows us to see how it can help augment more human things like abstraction, breaking rules, judgment, listening, and storytelling [8]. AuI, not AI, is the future. Therefore, businesses should be focusing on AuI instead of AI.

1.4 Edge Computing

Over the past few years, the internet of things (IoT) has risen from being just another futuristic buzzword to a tool having great utility and business value in the present. The internet is rife with real-life examples IoT applications; and from what we have seen and what we know about the technology, such cases are just scratching the surface of potential IoT applications. But just like any other form of technology, IoT comes with its own set of obstacles, or at least areas, that can be further enhanced. For starters, as IoT networks spread across and cover wide areas by incorporating a growing multitude of devices, the sheer volume of the data collected will require resource-intensive processing devices and high-capacity data centers. Similar problems arising due to the centralized, integrated nature of the technology can be eliminated by distributing the control and processing power of the IoT network toward the edge – the points where data is actually gathered, and where action, or rather reaction, is generally required.

1.5 Preparing for the IoT-Driven Future

The existing applications of IoT are already providing us with the evidence for a densely interconnected future, where every device will be able to communicate with every other device, creating an intricate web of information in and around our daily lives. These devices will be able to incessantly gather information through a myriad of sensors, process information through complex algorithms running on centralized servers, and effect changes using actuating endpoints. From agriculture to manufacturing, and health care to entertainment, every industry is set to see massive transformation driven by IoT.

Although the ability of IoT systems to execute and initiate responsive action will be transformational enough, the real revolution, as it were, would be brought about by the essentially limitless cornucopia of data that will be generated due to the unbridled proliferation of sensors and other data gathering IoT endpoints. In fact, this IoT data will prove to be the real wealth for the businesses using the technology, as structured data in unprecedented quantities can be captured and analyzed to gain deeper insights into the market and also into organizations and business processes. The increased volume of data gathered will enable businesses to take even more effective action, driving operational excellence. However, gathering and processing such vast amounts of data would require high-capacity storage, communication, and computational infrastructure. Even though advances in communications technology, such as the mainstream adoption of 5G, can catalyze IoT innovation and implementation, newer ways of making IoT more effective and efficient are still required. And one of the most promising solutions for enabling IoT to realize its potential is edge computing.

1.6 Defining the IoT's Edge

Edge computing refers to the installation and use of computational and storage capabilities closer to the edge, i.e., the endpoints where the data is gathered or where an immediate response is required. IoT systems can be composed of a large number and multiple types of endpoints connected to a centralized, often remotely located, data center. These endpoints include, but are not limited to: Computing devices used by employees that can be used to gather data; hand-held devices like smartphones and tablets that continuously generate data with use; sensors and sensor-based devices that gather data like temperature, radiation, current, footfall, and inventory levels; and actuators that can perform actions like operating switches, valves, motors, and transducers to control process parameters.

Edge computing in IoT implies having autonomous systems of devices at these endpoints (or the edge) that simultaneously gather information and respond to the information without having to communicate with a remotely constructed data center. Instead of having remote data centers and computational servers, the processing of data can be done right where the data is collected, eliminating the need for constant connectivity to centralized control systems and the problems inherently associated with such setups. For instance, a software company that sells cloud-based mobile applications can have cloud servers based in multiple locations closer to users instead of in a single location that may lead to undesirable latency and a single point of failure in case of any mishap. If the centralized servers failed due to some reason, all application users would lose their data and access to services at once. Additionally, the servers would also have to deal with heavy traffic, causing latency and inefficiency. On the contrary, a decentralized system would ensure that all the data pertinent to specific users would be hosted in the closest data center than among multiple ones, minimizing latency and limiting the impact of any potential failure. In addition to solving inherent IoT problems, the incorporation of edge computing into IoT is increasingly being seen as a necessity as it enhances the network in terms of functionality and performance.

1.7 Edge Computing Builds Privacy and Security

In the recent era, new technologies are emerging day by day causing multiple issues to pop up. Once they get resolved, many new problems arise. This is especially true when considering edge computing, where multiple IoT devices are deployed which are not very secure as they enable a centralized cloud platform. In advanced technological infrastructure, IoT devices are increasing enormously where security raises a potential concern and makes the devices to operate in a secured manner. To make the devices secure, an encryption strategy and authentication method must be adopted. The impact of data reliability also plays a significant role where different devices, which have different behaviors in terms of connectivity, processing time, etc. are needed. While processing these devices, data redundancy and failure management need to be ensured so that data can be delivered properly.

Generally, edge-based cloud computing is developing rapidly and proving its demand in the field of computing. Even though it is proving its ability to make edge-based devices powerful, there is a lack in device performance and need for reduced latency while it is integrated with machine learning, data analytics, etc. To compromise the above challenges, security and privacy play an important role while driving toward edge computing. While exposing millions of edge computing devices and IoT-based devices, it is clear that security plays a vital role in edge-based IoT computing.

1.8 Benefits of Edge Computing

Organizations using edge computing to power their IoT systems can minimize the latency of their network, i.e., they can minimize the time for response between client and server devices. Since the data centers are closer to the endpoints, there is no need for data to travel to and from the distant centralized systems. And as the edge storage and control systems are only required to handle the data from the few endpoints they are linked to, bandwidth issues seldom slow down the flow of data. Since IoT systems require high-speed information transfer to function with maximum efficacy, edge computing can significantly boost organizational performance.

Another benefit of decentralizing IoT with edge computing is providing data security. A centralized data repository is prone to hacks that aim to destroy, steal, or leak sensitive data, and such attacks can lead to the loss of valuable data. Conversely, distributing critical data across an IoT network and sitting on edge devices can limit the loss of data. Additionally, it can also help in compliance with data privacy rules, such as the General Data Protection Regulation (GDPR), since data is only stored in devices or subsystems that would use that data. For instance, a multinational corporation can use edge devices to store customer data on local devices that are closer to where the customers are, instead of storing the data in an overseas repository. The data need not be stored in locations where irrelevant personnel may have access to it. Cloud costs will also be minimized as most data will be stored on the edge devices, instead of on centralized cloud servers. Additionally, the cost of maintaining high-capacity, long-range networks will be reduced as bandwidth requirements continue to diminish. It is easy to see now why any discussion on IoT should always include the exploration of edge computing as a key enabler. Edge computing, more than a technology, is a design framework of sorts that would redefine the way IoT systems are built and the way they function. Although the combination of other solutions will also be needed to expedite the widespread adoption of IoT, edge computing might just prove to be the chief catalyst in the process.

1.9 Applications

AuI can be applied to many processes (e.g., business sales, decision making, identification verification, logistics planning, and remote assistance) and many industries (e.g., health care, journalism, insurance underwriting, manufacturing, finance, real estate, military, and legal).

- *Business sales*: The merger of humans and machines is critical to advancing modern business enterprise. Departmental heads, team leads, and vice presidents often identify business opportunities to apply AuI. Although AuI systems can be placed anywhere in a business organization, it is mostly used for business sales. Integrating intelligent capabilities into your sales department is a smart investment in your salespeople. AuI takes the complex sales tasks off the salespersons, enabling them to focus on meeting the needs of their customers. AI is capable of analyzing sales data and translate the data into action [9]. The range of business problems to which AuI applies continues to expand at a rapid pace.

- *Decision making*: If properly applied, AuI can provide feedback and insights that enhance decision-making. It is the ability of a manager to leverage AI and collective intelligence for every decision.

- *Logistics planning:* A logistics planner requires a lot of knowledge and experience about what works and what does not work in the industry. AuI comes in when AI is used to deal with high-skilled logistics planners. To improve logistics planning, companies should use AuI, which combines inputs from human planners with AI technology [12].

- *Health care:* Patients and health care practitioners face enormous challenges, such as rapidly aging populations, shortage of physicians, and high costs of care. This is partly illustrated by the quadruple aims of health care shown in Figure 1.4 [10]. The primary role for AI is augmentation of the intelligence and skills of care givers. AuI is not designed to replace health care practitioners but to enhance human intelligence and the physician–patient relationship [11].

- *Military*: The military deems a combination of humans and machines as necessary to cope with the complexity of information decoding and maximize the military ability to create, exploit, and adapt. AuI will harness the best of our soldiers and technology to meet future challenges [13–14].

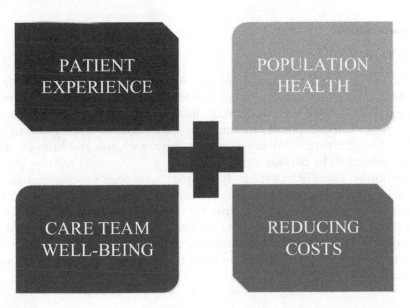

FIGURE 1.4
The quadruple aims of health care.

1.10 Challenges

AuI, the synergy between humans and AI, exploits the positive aspects of human and AI-generated reasoning. It is designed to play an assistive role to enhance human intelligence and to improve the quality and efficiency of workers. It is becoming an effective weapon to address the current blemishes of medicine which include poor predictive power, therapeutic errors, and inefficient hospital workflows [15]. AuI tools are used to make smarter decisions and save time on repetitive tasks. AuI is better suited for organizations than AI as it has a quicker return on investment and better utilizes the assets already in place. There are opportunities that will arise from "augmented" workplaces and life in general – it has the potential to transform social services organizations from case management to care management. Any field that requires both humans and technology will definitely be changed by AuI.

The idea that intelligence can be automated, replacing millions of humans in highly repeatable tasks, has received a great deal of attention. For the doomsayers, AI-augmented workplace implies an impending mass unemployment. The fears are essentially groundless. It appears that jobs will disappear rather than being automated. If machines can replace humans, we should be free to do what humans do best – create. For more information on AuI, one should consult the book in [16].

1.11 Conclusion

Intelligent machines (or computers) are efficient in numerical computation, information retrieval, and statistical reasoning, and have almost unlimited storage. They can see, hear, and speak multiple languages. They have become the intimate companions of humans; they are profoundly changing our lives and shaping the future. We are in an era where machines and humans working together will be pervasive and efficient.

The human brain is an extremely complex, advanced processing unit that can solve problems which computers simply cannot. No matter how intelligent machines may be, they are cannot completely replace humans. So, there is nothing to fear. We only need to prepare. In the foreseeable future, AI will be augmenting our capabilities, allowing us to do more in less time. Rather than focusing on AI, we should consider the possibilities of how businesses can take advantage of AuI.

References

1. M. Sharma, "Augmented intelligence: A way for helping universities to make smarter decisions," in Rathore, V.S., Worring, M., Mishra, D.K., Joshi, A., Maheshwari, S. (eds.), *Emerging Trends in Expert Applications and Security.* Singapore: Springer, 2019, pp. 89–95.
2. "What is augmented intelligence?" www.quora.com/What-is-augmented-int elligence, Abhinav Jain, 2017.
3. "The future of decision-making: Augmented intelligence.", Insight, Genpact, 2020 www.genpact.com/insight/point-of-view/the-future-of-decision-maki ng-augmented-intelligence
4. "Augmented analytics plays an important role in transform big data to smart data", Frost & Sullivan, March 13, 2019, https://timestech.in/augmented-anal ytics-plays-a-important-role-in-transform-big-data-to-smart-data/
5. M. N. O. Sadiku, "Artificial intelligence," *IEEE Potentials*, May 1989, pp. 35–39.
6. "Applications of artificial intelligence (AI)," July 2017, Beginners, Blog, Computer System Engineering, www.mechlectures.com/applications-artificial-intelligen ce-ai/
7. A. Lui and G. W. Lamb, "Artificial intelligence and augmented intelligence collaboration: Regaining trust and confidence in the financial sector," *Information & Communications Technology Law*, vol. 27, no. 3, 2018, pp. 267–283.
8. "What is augmented intelligence?" March 2018, Aisling McCarthy, media update. www.mediaupdate.co.za/media/143606/what-is-augmented-intelligence
9. S. Akhrif, "Augmented intelligence and the future of selling," https://uib.ai/ augmented-intelligence-future-selling/2018
10. F. Pennic, "AI in healthcare: Enhancing risk models to predict the future cost of care," September 2019, https://hitconsultant.net/2019/07/09/ai-in-healthcare-e nhancing-risk-models-to-predict-the-future-cost-of-care/#.XZPTNndFwc8
11. "Health imaging: Augmented intelligence the next frontier," Anjum M. Ahmed, AGFA HealthCare, 2018, www.healthcareitnews.com/news/augmented-intel ligence-next-frontier-health-imaging
12. "Augmented intelligence for logistics planning,", Asparuh Koev, Transmetrics, 2019, https://transmetrics.eu/blog/augmented-intelligence-for-logistics-plan ning/
13. W. L. Koh, Joyce Kaliappan, Mark Rice, Keng-Teck Ma, Hong Huei Tay, Wah Pheow Tan, "Preliminary investigation of augmented intelligence for remote assistance using a wearable display," *Proceedings of the 2017 IEEE Region 10 Conference*, Malaysia, November 2017, pp. 2093–2098.
14. "Augmented intelligence," Plan Jericho, 2019, www.airforce.gov.au/our-missio n/plan-jericho/augmented-intelligence
15. M. Bhandari and M. Reddiboina, "Augmented intelligence: A synergy between man and the machine," *Indian Journal of Urology*, vol. 35, no. 2, April–June 2019, pp. 89–91.
16. A. Bates, *Augmented Mind: AI, Humans and the Superhuman Revolution.* Neocortex Ventures LLC, 2019.

2

Edge Computing and Embedded Storage

Ramkumar Jayaraman, M. Baskar, and B. Amutha

CONTENTS

ABSTRACT Recent technological developments focus on improvements to embedded devices and the ability for them to provide the technology's service in a simple, effective and cost-cutting way with the communication of sensors and other devices. While claiming the work of embedded storage, devices like sensors collect the information and send the data to other ends via the internet, such as the internet of things (IoT) gateway. Regarding storage, cloud computing and cloud-data centers play a significant role. As cloud computing has some limitations relating to cost, latency, and connection it provides remote connection and management for operational technology (OT), which results in the business choice of cloud storage. Even though the user's storage needs are surplus, some applications with various parameters like latency, cost, and energy efficiency are required to monitor the storage components periodically and improve the flow of data via the cloud. Based on the cloud-application usage, the response time of data centers is getting reduced with technology improvement.

Even though many issues still exist, the security and connectivity issue is emerging when, for example, a vehicle integrates with the sensors of intelligent devices while sending data to check the position and status of the vehicle. Those devices should be managed and provide a solution to the issues of security and connectivity. As the above issues provide the solution by increasing the number of intelligent devices, there is still one issue – cost. Whenever the number of devices keeps on increasing to provide various activities, monitoring and maintaining those devices becomes expensive. For this reason, data is transferred through data centers which filter the unwanted data and is then loaded into the data center to get transferred. As we have discussed, some of the significant problems which arise are based on edge computing with the integration of embedded storage devices. In this chapter, first we discuss edge computing and edge computing patterns followed to work with various applications related to IoT and networks, etc. Then, we discuss how edge computing helps in building the security and privacy in the application platform, and how the present works carried out so far to mitigate the security breaches prevailed in various fields over edge computing. Finally, we point out the open challenges that still exist and how it provides opportunities for further research in the future.

2.1 Introduction

In the recent technological era, rapid developments in 5G wireless networks concerning fast data rates and IoT-integrated devices when data are created are based on edge computing technology [1]. In the above case, edge technology helps in transforming and handling the data to get processed and delivered to an enormous number of IoT devices around the world. As we all know, IoT devices are growing enormously and are integrated with many new applications, which require real-time data transfer and power consumption based on edge computing technology [2–4]. In recent years, we have moved to edge computing, which has the ability to secure the process of computing and then store the information analysis at the end of the process [5]. Nowadays, we are moving to the creation of new real-time applications like self-driving cars, robotics, and artificial intelligence (AI) [6 and 7], video processing, and data analytics. The goal of the edge computing is to obtain improved bandwidth as it requires longer distance and time consumption to process data through IoT devices [8].

Generally, edge computing can process data based on the distributed topology and is currently used in several real-time applications and process the information by people (or) other IoT devices [9]. The edge computing provides management functionalities like data analytics, data filtering, data aggregation, routing the data and maintain device management [10] and [11]. Hence, edge computing provides standard management functionalities and supports advanced functionalities to process the data, which reduces the latency with proper connection and managing end to end infrastructure as represented in Figure 2.1. As several manufacturing companies are working under edge computing, Euro Tech manufacturing company is working for the past 25 years to manufacture shrunken computers with the integration of several market environment [12]. Generally, edge computing gets it's evolved based on the enormous growth of IoT devices when the data are transfer to the cloud or from the cloud and IoT devices are generated data enormously during the process of edge operation [13].

2.2.1 The Working of Edge Computing

Generally, edge computing takes data collected from IoT devices which is then analyzed and processed, and then sent to the data center or cloud to get processed. At the earlier stage, the computing edge makes all the necessary

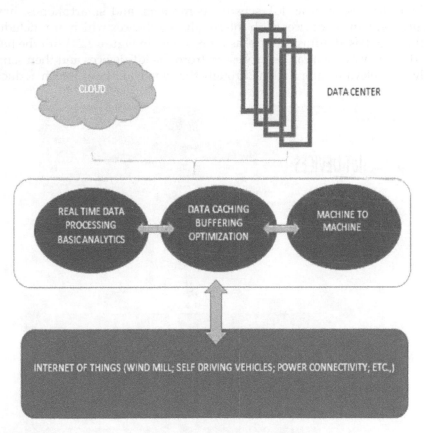

FIGURE 2.1
Edge computing.

data for computation and gathers it. Computer edging brings data close to the IoT devices instead of making that data move to the central data center, which may be located many miles away from them [14]. In this process, latency issues are avoided, and thus the performance of an application is improved. In addition to the latency reduction, cost is also reduced when the data is processed locally through a centralized process or cloud-based location.

IoT devices will be manufactured and data received from them will be monitored remotely from a centralized location. In the scenario of IoT devices, when a single device transmits data through a network, it is easy to transmit, but when multiple devices transmit data at the same time, performance, connectivity, and other problems arise [15]. In this scenario, data quality reduces and latency suffers from an increase in bandwidth cost. The problems get rectified based on hardware and services, which process the data locally and store based on multiple systems.

Edge devices, such as IoT sensors, computers, and smartphones, have security-enhanced devices with internet-based devices which are included in the edge-based infrastructure, as represented in Figure 2.2. With the help of edge gateway, data can be processed from the IoT devices and then sends only the relevant information through the cloud location, which reduces

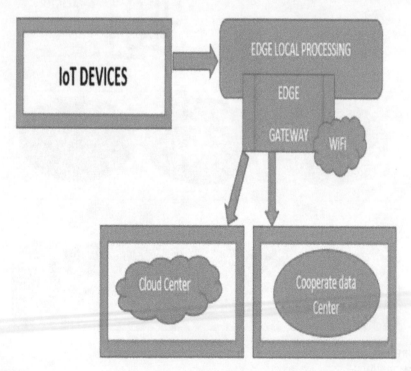

FIGURE 2.2
Edge computing architecture.

the bandwidth. After data gets processed, data can also be sent back to IoT devices for real-time applications. Thus, the edge computing process takes place with specific components including IoT devices, local edge processing, edge gateway, cloud-based location, and data center. Based on edge computing components, edge technology architecture can be deployed in various companies to save cost. As many companies are concentrating on reducing bandwidth cost and increasing performance, they are adopting cloud platforms in real-time applications [15].

As edge computing deploys in various applications, it provides certain benefits relating to the ability to process the data and allow for faster storage of data, which helps to make the edge technology in real-time applications to perform effectively. As we all know, before the deployment of edge technology, facial recognition processed the face through a certain algorithm, which takes quite some time for data processing. The edge computing model is considered in facial processing by applying the algorithm through the local server, i.e., edge gateway [16]. We can also apply the same edge technology in other applications like virtual, augmented, and mixed reality; self-driving cars as it requires faster processing speed; and smart cities as it requires multiple IoT devices to be deployed and transfer an enormous amount of data for processing [17]. During the scenario of multiple IoT devices, improved connectivity through edge technology gets adopted and improves the edge infrastructure, which requires improved data processing and data storage in the new environment.

2.2.2 Edge Computing Builds Privacy and Security

In today's era, new technologies are emerging day by day and multiple issues are surfacing. When these issues get solved, they may result in new problems. This is especially true when considering edge computing, where multiple IoT devices are deployed which is not very secure as it enables a centralized cloud platform [18]. As we all know, the number of IoT devices is getting increased in this digital age, where security is a potential issue and where IoT devices are operated in an unsecured manner. Encryption strategy and authentication methods have to be adopted to make the IoT devices secured. The impact of data reliability also plays a significant role when applying different devices, as these have different behaviors of connectivity, processing time, etc. While processing the different devices, data redundancy and failure management among different devices need to be managed so that data is adequately delivered [19].

Generally, edge-based cloud computing is developing rapidly and proving its demand in the field of edge computing. Even though it is proving its status to make edge-based devices powerful, there is a lack of device performance and the need to reduce latency while it integrates with machine learning, data analytics, etc. Security and privacy play an important role while driving toward edge computing to overcome the above challenges. While exposing millions of edge computing devices and IoT-based devices, security plays a vital role in edge computing-based IoT [20].

From Figure 2.3, we infer that the public cloud provides several services through the internet, where users can provide enhanced security at a higher level. Nowadays, edge computing and IoT solutions are providing specific solutions based on differentiation in data traffic flow. Recent attacks among the public cloud-based edge is denial of service (DoS) and distributed denial of service (DDoS) attacks [21]. As we know, DoS and DDoS attacks affect the edge of the security and traffic flow among the data in the application. To solve certain problems, edge computing provides a separate edge cloud while providing services to each application. This edge cloud helps to avoid attacks when it serves each application and helps to provide intrusion detection and prevention systems against DoS and DDoS attacks [22] when it offers the services of computation, storage, and networking management. By providing this edge cloud, which acts as a deep packet inspection, it serves to prevent the data from being infected with malware. Following the divide and conquer model, the edge cloud is divided into several segments, which is represented in Figure 2.4. It divides the larger sets of data into multiple subsets where these subsets are offered to each application via segmented edge clouds to isolate the massive attacks into multiple subattacks to get managed by the edge cloud.

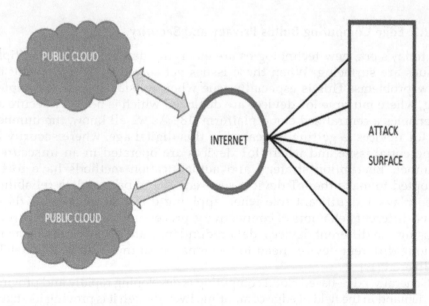

FIGURE 2.3
Edge computing with attacks based on the public cloud.

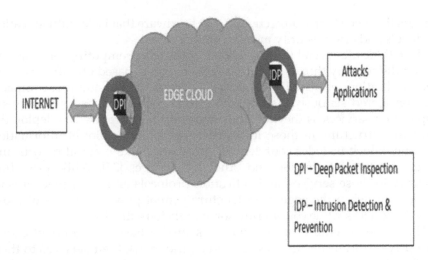

FIGURE 2.4
Individual segmented edge cloud.

2.2 Related Works Based on Security Threats

Even though edge computing considers the problems relating to security against DoS and DDoS attacks, here we discuss some other challenges relating to security in edge computing [23] and we will categorize the security threats in edge computing. Despite several challenges prevailing in edge computing based on infrastructure and providers, security is one of the significant challenges existing in recent days.

In this section, we discuss the several challenges of security relating to the edge paradigm and its requirements. First, we discuss the significance and requirements of security in edge computing.

In creating the edge computing infrastructure, security is a critical problem facing users. First, while creating infrastructure, we need to enable several technologies like networking technologies, distributed and centralized environments, and software-defined and virtualized structures. We cannot protect all enabled technologies, so we use edge computing to coordinate all the enablers and provide security. Even though the edge computing coordinates all the enablers, it is difficult to provide security as a whole. Next, edge computing combines all the enablers into one and provides security is to an individual rather than the coordinated one [24]. So edge computing is an integrated cloud, which includes data centers and exchanging the services from global to local. In the above context also, security plays a challenging role. When edge computing integrates with the data center, it contains multiple servers that need better protection to that hardware that lacks authentication protocols and other security mechanisms.

Next, there are several features added in the edge computing paradigm in which the security threats need to be concentrated, based on infrastructure and application services. Finally, attacks are carried out and those impacts are considered seriously. As we know, edge computing integrates several application services as they are enormous and they are used to deploy the edge infrastructure. In those infrastructures, there is a lot of information shifting, which includes confidential information like medical reports and other sensitive information and other IoT services [25]. While providing security to those services, authentication protocols and other mechanisms are lacking when the edge infrastructure cannot provide benefits of those specific services while considering some malicious attacks.

These are several scenarios where security based on authentication and privacy protocols are lacking and cannot mark their services to their levels.

2.3 Threat Background

In this section, we discuss the importance of security related to edge computing, where some threats and malware are marching toward these edge computations and causing damage. Based on the analysis of security importance, edge computing provides both an effective security mechanism and security to the whole environment.

Before analyzing the attack, check how it occurs and the deployment/ computation of the security in the edge technology environment. Mainly, an analysis of the security perimeter level is needed to mitigate the threats and to what extend the threat could breach the security. In the edge computing environment, the entire system is not controlled by the single user as it applies to a cloud distributed platform. In cloud computing, data centers are providing services without concentrating on a centralized entity and here, the services include infrastructure networks and services which serve as data centers and servers, network virtualization, and user devices [26]. The above services are provided not based on a centralized manner but based on a distributed manner, with cooperation and coordination among those services offered in edge computing, as represented in Figure 2.5. By having those services, we apply the principle to anything, anytime, and to all the services. It also applies to various other scenarios, like machine learning and the IoT.

As we mentioned before, edge computing uses the "anytime and any-where" principle, which allows for various attacks and malware to occur at any time. Generally, cloud computing acts as data centers that provide

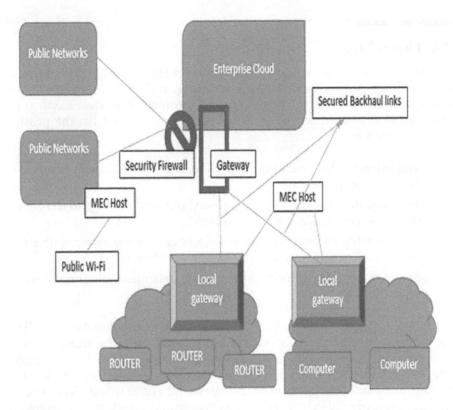

FIGURE 2.5
Edge computing with multiple services.

services based on Mobile Edge Computing (MEC) host and nodes as local entities, i.e., closed environment as services provided within the particular building premises. Apart from the edge computing concept, the process of the virtual machine comes into the picture where systems and servers are included and moved to other infrastructure without taking the physical movement into control [27]. By using the virtual concept, we have two points as a double-edged sword. The first is the virtual reduction to the attack in the local environment and the second, it has a remote edge server which provides services and monitors all the services in a particular geographical location.

Based on edge computing, we have a lack of a global perimeter while considering the virtual machine. Several attacks are possible, which concentrate based on edge prototype. When particular attacks happen, it should not attack all the elements in the particular prototype. On the other side, when the attacks happen to all the elements in the edge model, then elements include devices, virtual machines, and servers. In the current scenario, when a particular malware activity happens and it takes control of all the elements and then deploys its basis activity.

2.4 Threat Model

Till now, we have discussed the significance of the attacks and the fact that they are potentially related to edge computing, and now we are able to analyze the threats presented. For an analysis of threats, below we mention the essential assets based on the edge computing model and list the possible threats to each asset.

Asset: Infrastructure network → DoS, man-in-the-middle attack, and gateway rogue (threats).

Asset: Edge data center → rogue gateway, physical damage to the data center/servers, privacy leakage (threats).

Asset: Infrastructure virtualization → DoS, virtual machine manipulation, resource misuse, and privacy leakage (threats).

Asset: user device → information injection and manipulating the services (threats).

When a particular threat affects the edge model, then the threat affects all the components/assets in the edge ecosystem. In addition to the above information, we need to analyze the context of the edge model in both a centralized and a distributed manner and consider the additional service like location management and interoperability [28]. As we all know, threats have a major impact when they affect the edge data center and attack the geographical management. Thus when the attacks happen frequently, new threats continue to grow.

Thus, we discuss the assets and threats that arise in edge computing briefly below.

2.4.1 Infrastructure Network

As we know, some services are applied based on the network communication where all the devices/elements used in the edge model are coordinated based on the wireless network and mobile networks. Based on the network, we discuss some of the possible threats that occur, as mentioned below.

A. Denial of Service

When a single system attacks multiple servers, it is termed as DoS. When all communication systems attack multiple servers, it is termed as DDoS [29], as represented in Figure 2.6. The possibility of these DDoS and DoS attacks is restricted but when a particular attack happens on the edge model, several indications show that the edge computing network is being affected. When this particular DoS attacks the infrastructure of the edge core, which is nothing but

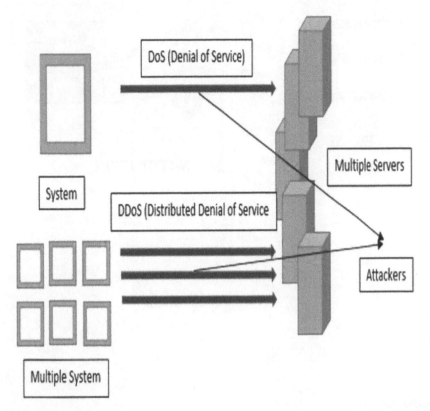

FIGURE 2.6
DoS versus DDoS attacks.

the data center, it won't affect the edge protocols and services offered by the edge computing as it is independent behavior.

B. **Man-in-the-Middle Attack**

Man-in-the-middle attacks happen when the particular attacker takes control of the communication in the network, and is then able to inject or perform malicious activity, as represented in Figure 2.7. We call this activity "injection traffic" or "eavesdropping". There may be a possibility of constructing heterogeneous networks like Wireless Fidelity (WiFi), Wireless interoperability for Microwave Access (WiMAX), and Long Term Evolution (LTE), [30] which are interconnected to form a particular network. If these kinds of network receive a man-in-the-middle attack, control of all the activity performed will be affected in all the elements in the network, and it is more dangerous and hazardous.

C. **Rogue Gateway**

In this edge computing scenario, all the user devices can also become the user participants who can also be involved in network

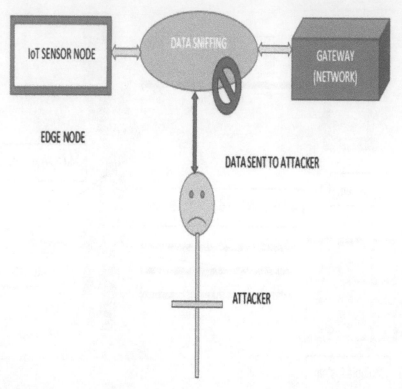

FIGURE 2.7
Man-in-the-middle attack in edge computing.

communication. In these situations, when a malicious attack happens, malware becomes the user device and it attacks in the same way as a man-in-the-middle attack [31].

2.4.2 Infrastructure as Service

In this infrastructure, edge data centers deploy the servers virtually and based on that, the services offered. Edge data center attacks are significant; these attacks are discussed below based on infrastructure as a service.

A. Damage Physically

These attacks happen on the physical components of the edge component, where the service offered is not secured or protected against the malware activity. When considering a small organization with a device connected with clusters, attackers will destroy all the physical components in that organization. Challenges on those attacks are limited as the devices/services are associated with location-enabled devices.

B. **Leaking in Privacy**

These attacks are limited, but when they occur, the flow of data between the edge data centers is attacked. The attack takes place during the flow of data and it modifies/extracts the information from the communication network. It applies to distributed storage and virtual migration. During security inconsistency, attackers copy or extract the most confidential data while the data moves between the edge data centers [32].

C. **Improving Privilege**

While attacking the edge data center, the attacker takes control of the various services offered by edge computing and such activities are mostly carried out by well-equipped limited security trainers. Due to improper network/system configuration with low security, this paves the way for the attacker to modify and extract data.

D. **Manipulating the Service**

In the service manipulation, attacks take control through communication in the network by exploiting the privilege among the edge data center [33]. During the attack, the attacker takes control of the network. The privilege control is exploited by the trusted employee who is working as an administrator for access. During the above scenarios, attacks like DoS and data tampering occurs with communication in the network.

E. **Reprobating the Data Center**

In this situation, the attacker attacks the edge data center and makes malicious information to be penetrated in the network [34], as represented in Figure 2.8. While it penetrates the network, it takes some privileges such as:

1. First, it takes control of all the services offered by edge computing and makes it possible based on the geographical location information.

2. After taking all the service from the above step, it transfers all the data communication information to the X data center, i.e., unauthorized party who makes the malicious activity.

3. Finally, when it is sent to the X data center, it makes some modifications to the original communication through third parties, as the original communications are performed in the virtual machine and services are performed on remote.

2.4.3 Infrastructure as Virtualization

All edge data centers provide several services, but in these scenarios, all the services are offered based on virtualization machines through cloud

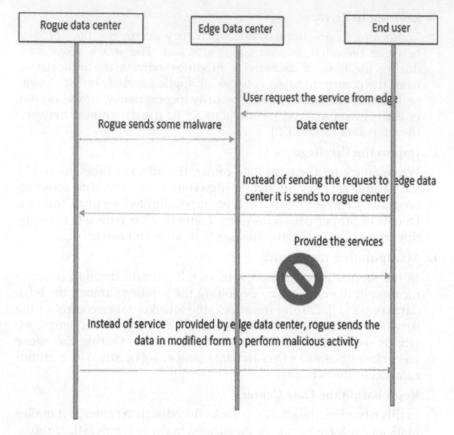

FIGURE 2.8
The reprobating situation in data centers.

platforms [35]. Even though the virtualization avoids or mitigates the security breaches, there are some security problems that may occur. These are listed below.

A. **Denial of Service**

During the DoS, the malware injects itself into the machine virtually, which results in improper activity among various resources like storage and network computation [30 and 29]. When trying to perform the above activity, the attacker targets the machine by making use of the resources available on the cloud platform.

B. **Information Abuse**

This occurs when the virtual machine with malware tries to perform an attack on the source host and creates malicious activities in the network. As we know, the virtual machine is not performing on the source host; instead, it performs on the local host or remote host

[36]. So, the malware makes the activity to perform on the local host and tries to crack the identification information (like the password) to identify and attack the source host.

C. **Outflow and Malicious Problems During Information Privacy**

When the virtual machine itself acts as an infrastructure, it will not process the information at the edge data center. In fact, it will get identified and it's location information is not visible. It may be located anywhere through the local or remote host and they are connected with the edge data center through application interface as it knows where the source host resides. So, when malicious activity happens, it attacks the application interface and knows all the information about the source edge data center.

If the malicious activity is identified and performed in any one of the virtual machines, the virtual machine takes advantage and moves one virtual machine to another. Because the malicious virtual host attacks, they take all the privilege and control. So the virtual machine moves and connects to another host to continue its activity [37].

D. **Manipulating the Virtual Machine**

As we know, the host data center controls all the computational activity so it may also be affected by the malware. When the malware attacks the host, it takes control of the information extraction and the computational activity performed and executed. When this activity takes place, edge data center security is compromised.

2.4.4 User Devices

As we know, edge computing provides some services and performs the activity based on the devices, and plays an essential role in a particular environment. The system is formulated based on distribution, which offers certain services by the system and is identified/controlled by the devices, which establishes in the network [38]. Users monitor and control the devices which are able to establish the services for specific applications, but when the particular users are rogue, the services provide disturbance and lead to the security breach and influence the attack.

A. **Unwanted Information Penetration**

When the edge data center offers the services, which get attacked by the malware, then the information produced by the services is disturbed and becomes useless, and is then sent back to the system. For example, we consider the IoT environment, which provides several sensors and provides some information like temperature and humidity to the system [39]. But when the malware attacks the sensor's services, then that information obtained by the sensors gets distributed and become useless.

B. **Manipulating the Services Based on the User Device**

The edge computing provides several services, which makes the machine to perform virtually when it is placed at the edge data center. Those devices are made as cluster and that role plays a significant role and they establish distributed fashion. When it is placed in a distributed fashion among the devices, the malware activity affects the user devices and can take control of the services and is able to manipulate the information in the network.

From the above discussion, we have identified some security breaches among those elements, such as infrastructure as services, edge center, infrastructure as the core, infrastructure as virtualization [40], and user devices. Then the above elements get breached by several attacks like DoS, DDoS and data tampering.

2.5 Open Challenges Relating to Security in Edge Computing

The infrastructure as services, edge center, infrastructure as the core, infrastructure as virtualization, and user devices. Elements get breached by several attacks like DoS and data tempering. As can be seen, there are many security issues and there is room for research work to carry out. The open challenges relating to edge computing on security breaches are listed below. Here we also list some of the other challenges related to QoS in edge computing.

a. IoT industrial-based edge computing.

b. IoT-based smart cities and infrastructure through edge computing.

c. Customized data collection and analysis in hospitals through edge data centers.

d. Computing hardware constraints in the edge environment.

e. Accessing the constraints relating to operation through edge computing applications.

f. Managing and deploying remote hosts through virtualization in edge computation.

g. Secure communication through edge data centers.

h. Ensuring privacy and data integrity in edge computing.

i. Trust maintenance based on the rogue system in edge computing.

2.6 Summary

We have elaborated on the security breaches and QoS maintenance in edge computing. We have identified some security breaches in certain elements such as infrastructure as services, edge center, infrastructure as the core, infrastructure as virtualization, and user devices. Those elements get breached by several attacks like DoS and data tampering We also outlined some of the significant problems that arise based on edge computing with the integration of embedded storage devices and infrastructure based on security enablers. Finally, we discussed edge computing, its components, and its challenges and solutions and the fact that open problems still exist.

References

1. Wazir ZadaKhan, Ejaz Ahmed, Saqib Hakak, Ibrar Yaqoob, Arif Ahmed, "Edge Computing: A Survey," *Future Generation Computer Systems*, Vol. 96, pp. 219–235, August 2019.
2. Maurizio Capra, Riccardo Peloso, Guido Masera, Massimo Ruo Roch, Maurizio Martina, "Edge Computing: A Survey on the Hardware Requirements in the Internet of Things World," *Future Internet, MDPI*, Vol. 11, Issue 4, pp. 1–25, April 2019.
3. Fang Liu, Guoming Tang, Youhuizi Li, Zhiping Cai, Xingzhou Zhang, Tongqing Zhou, "A Survey on Edge Computing Systems and Tools," *Proceedings of the IEEE*, Vol. 107, Issue 8, pp. 1537–1562, June 2019.
4. Urooj Yousuf Khan, Tariq Rahim Soomro, "Applications of IoT: Mobile Edge Computing Perspectives," in *Proc. of 12th International Conference on Mathematics, Actuarial Science, Computer Science and Statistics (MACS)*, Karachi, Pakistan, November 2018.
5. Cheol-Ho Hong, Blesson Varghese, "Resource Management in Fog/Edge Computing: A Survey on Architectures, Infrastructure, and Algorithms," *ACM Computing Surveys*, Vol. 52, Issue 5, pp. 97–137, September 2019.
6. Weisong Shi, George Pallis, Zhiwei Xu, "Edge Computing [Scanning the Issue]," *Proceedings of the IEEE*, Vol. 107, Issue 8, pp. 1474–1481, August 2018.
7. Charif Mahmoudi, Fabrice Mourlin, Abdella Battou, "Formal Definition of Edge Computing: An Emphasis on Mobile Cloud and IoT Composition," in *Proc. of 2018 Third International Conference on Fog and Mobile Edge Computing (FMEC)*, Barcelona, Spain, April 2018.
8. Saksham Mittal, Neelam Negi, Rahul Chauhan, "Integration of Edge Computing with Cloud Computing," in *Proc. of International Conference on Emerging Trends in Computing and Communication Technologies (ICETCCT)*, Dehradun, India, November 2017.

9. Jiale Zhang, Bing Chen, Yanchao Zhao, Xiang Cheng, Feng Hu, "Data Security and Privacy-Preserving in Edge Computing Paradigm: Survey and Open Issues," *IEEE Access*, Vol. 6, pp. 18209–18237, March 2018.

10. Najmul Hassan, Kok-Lim Alvin Yau, Celimuge Wu, "Edge Computing in 5G: A Review," *IEEE Access*, Vol. 7, pp. 127276–127289, August 2019.

11. Najmul Hassan, Saira Gillani, Ejaz Ahmed, Ibrar Yaqoob, Muhammad Imran, "The Role of Edge Computing in Internet of Things," *IEEE Communications Magazine*, Vol. 56, Issue 11, pp. 110–115, November 2018.

12. Zhuang Chen, Qian He, Lei Liu, Dapeng Lan, Hwei-Ming Chung, Zhifei Mao, "An Artificial Intelligence Perspective on Mobile Edge Computing," in *Proc. of IEEE International Conference on Smart Internet of Things (SmartIoT)*, Tianjin, China, August 2019.

13. G. Suganya, K. Kumar, "A Survey on Edge Computing Using IOT Systems," in *Proc. of 3rd International Conference on Communication and Electronics Systems (ICCES)*, Coimbatore, India, October 2018.

14. Kerem Aytaç, Ömer Korçak, "IoT Edge Computing in Quick Service Restaurants," in *Proc. of 16th International Symposium on Modeling and Optimization in Mobile, Ad Hoc, and Wireless Networks (WiOpt)*, Shanghai, China, May 2018.

15. T. Madhu Perkin, S. Mini, "Assignment of IoT Nodes to Edge Computing Devices in Internet of Things," in *Proc. of European Conference on Networks and Communications (EuCNC)*, Valencia, Spain, June 2019.

16. Kosisochukwu J. Madukwe, Ijeoma J. F. Ezika, Ogechukwu N. Iloanusi, "Leveraging Edge Analysis for Internet of Things Based Healthcare Solutions," in *Proc. of IEEE 3rd International Conference on Electro-Technology for National Development (NIGERCON)*, Owerri, Nigeria, November 2017.

17. Michael Schneider, Jason Rambach, Didier Stricker, "Augmented Reality Based on Edge Computing Using the Example of Remote Live Support," in *IEEE International Conference on Industrial Technology (ICIT)*, Toronto, ON, Canada, March 2017.

18. Melanie Heck, Janick Edinger, Dominik Schaefer, Christian Becker, "IoT Applications in Fog and Edge Computing: Where Are We and Where Are We Going?," in *Proc. of 27th International Conference on Computer Communication and Networks (ICCCN)*, Hangzhou, China, August 2018.

19. Kewei Sha, T. Andrew Yang, Wei Wei, Sadegh Davari, "A Survey of Edge Computing Based Designs for IoT Security," *Digital Communications and Networks*, Vol. 6, Issue. 2, pp. 195–202, May 2020.

20. Hokeun Kim, Edward A. Lee, Schahram Dustdar, "Creating a Resilient IoT with Edge Computing," *IEEE Computer*, Vol. 52, Issue 8, pp. 43–53, August 2019.

21. Deepali, Kriti Bhushan, "DDoS Attack Defense Framework for Cloud Using Fog Computing," in *Proc. of 2nd IEEE International Conference on Recent Trends in Electronics, Information & Communication Technology (RTEICT)*, Bangalore, India, May 2017.

22. Hanan Mustapha, Ahmed M Alghamdi, "DDoS Attacks on the Internet of Things and Their Prevention Methods," in *Proc. of the 2nd International Conference on Future Networks and Distributed Systems*, Vol. 4, pp. 1–5, June 2018.

23. Ruchi Vishwakarma, Ankit Kumar Jain, "A Survey of DDoS Attacking Techniques and Defence Mechanisms in the IoT Network," *Telecommunication Systems*, Vol. 23, pp. 3–25, July 2019.

24. Jianli Pan, James Mc Elhannon, "Future Edge Cloud and Edge Computing for Internet of Things Applications," *IEEE Internet of Things Journal*, Vol. 5, Issue 1, pp. 439–449, February 2018.
25. Kashif Bilal, Osman Khalid, Aiman Erbad, Samee U. Khan, "Potentials, Trends, and Prospects in Edge Technologies: Fog, Cloudlet, Mobile Edge, and Micro Data Centers," *Computer Networks*, Vol. 15, pp. 94–120, January 2018.
26. Roberto Morabito, Vittorio Cozzolino, Aaron Yi Ding, Nicklas Beijar, Jorg Ott, "Consolidate IoT Edge Computing with Lightweight Virtualization," *IEEE Network*, Vol. 32, Issue 1, pp. 102–111, February 2018.
27. Mahadev Satyanarayanan, "The Emergence of Edge Computing," *IEEE Computer*, Vol. 50, Issue 1, pp. 30–39, January 2017.
28. Liang Xiao, Xiaoyue Wan, Canhuang Dai, Xiaojiang Du, Xiang Chen, Mohsen Guizani, "Security in Mobile Edge Caching with Reinforcement Learning," *IEEE Wireless Communications*, Vol. 25, Issue 3, pp. 116–122, June 2018.
29. Qiao Yan, Wenyao Huang, Xupeng Luo, Qingxiang Gong, F. Richard Yu, "A Multi-Level DDoS Mitigation Framework for the Industrial Internet of Things," *IEEE Communications Magazine*, Vol. 56, Issue 2, pp. 30–36, February 2018.
30. L. Feinstein, D. Schnackenberg, R. Balupari, D. Kindred, "Statistical Approaches to DDoS Attack Detection and Response," in *Proc. of DARPA Information Survivability Conference and Exposition*, Washington, DC, USA, April 2003.
31. Shanhe Yi, Zhengrui Qin, Qun Li, "Security and Privacy Issues of Fog Computing: A Survey," in *Proc. of International Conference on Wireless Algorithms, Systems, and Applications WASA 2015: Wireless Algorithms, Systems, and Applications*, Qufu, China, pp. 685–695, August 2015.
32. Chang Liu, Yu Cao, Yan Luo, Guanling Chen, Vinod Vokkarane, Ma Yunsheng, Songqing Chen, Peng Hou, "A New Deep Learning-Based Food Recognition System for Dietary Assessment on an Edge Computing Service Infrastructure," *IEEE Transactions on Services Computing*, Vol. 11, Issue 2, pp. 249–261, April 2018.
33. Bastien Confais, Adrien Lebre, Benoît Parrein, "An Object Store Service for a Fog/Edge Computing Infrastructure Based on IPFS and a Scale-Out NAS," in *Proc. of IEEE 1st International Conference on Fog and Edge Computing (ICFEC)*, Madrid, Spain, August 2017.
34. Bastien Confais, Adrien Lebre, Benoît Parrein, "Performance Analysis of Object Store Systems in a Fog and Edge Computing Infrastructure," *Transactions on Large-Scale Data- and Knowledge-Centered Systems*, Vol. XXXIII, pp. 40–79, August 2017.
35. Roberto Morabito "Virtualization on Internet of Things Edge Devices with Container Technologies: A Performance Evaluation," *IEEE Access*, Vol. 5, pp. 8835–8850, May 2017.
36. Kai Han, Shengru Li, Shaofei Tang, Huibai Huang, Sicheng Zhao, Guilu Fu, Zuqing Zhu, "Application-Driven End-to-End Slicing: When Wireless Network Virtualization Orchestrates with NFV-Based Mobile Edge Computing," *IEEE Access*, Vol. 6, pp. 26567–26577, May 2018.
37. Flávio Ramalho, Augusto Neto, "Virtualization at the Network Edge: A Performance Comparison," in *Proc. of IEEE 17th International Symposium on a World of Wireless, Mobile and Multimedia Networks (WoWMoM)*, Coimbra, Portugal, July 2016.
38. Weisong Shi, Schahram Dustdar, "The Promise of Edge Computing," *IEEE Computer*, Vol. 49, Issue 5, pp. 78–81, May 2016.

39. Tuyen X. Tran, Abolfazl Hajisami, Parul Pandey, Dario Pompili, "Collaborative Mobile Edge Computing in 5G Networks: New Paradigms, Scenarios, and Challenges," *IEEE Communications Magazine*, Vol. 55, Issue 4, pp. 54–61, April 2017.

40. Xu Chen, Qian Shi, Lei Yang, Jie Xu, "Thrifty Edge: Resource-Efficient Edge Computing for Intelligent IoT Applications," *IEEE Network*, Vol. 32, Issue 1, pp. 61–65, February 2018.

3

Application of VANET to Avoid Pedestrian Collision in Automotive Vehicles

M. Dhana Lakshmi Bhavani and R. Murugan

CONTENTS

ABSTRACT Machines are playing a key role in most innovative industrial environments. These developments in terms of machinery can be improved by additional intelligence incorporated to the machines. Intelligent machines have the ability to help people in various scenarios of their day-to-day lives. For example, intelligent machines can be applied in various fields and situations, such as continuous monitoring and diagnosing patients, video surveillance, intelligent transportation system, communication between machines, and online education systems. In the near future, machinery will be the most integrated part of any company or organization for efficient and effective working. This chapter describes the modification to a special module for intelligent automobiles for safe driving on roads and also to avoid pedestrian collisions. Pedestrian protection is very important since there are many people who lose their lives in road accidents. An intelligent module should be developed to detect the presence of people crossing roads and then providing necessary alerts to the corresponding automobile in order to avoid collision and possible injury. One of the practical situations where the driver assistance system is compromised is in weather with heavy fog. In such scenarios, tracking the presence of pedestrians is a somewhat difficult task. To avoid this, we can introduce a specialized camera setup for clear vision information. The same pedestrian track information should be passed to the next adjacent vehicle that cannot view the presence of the approaching pedestrian. Such types of development improve safety measures and contributes to a comfortable lifestyle, even for physically challenged people in our society. This type of communication can be achieved by a special networking model called vehicular ad hoc network (VANET). These network models have no particular infrastructure and functionality. The features of automobile characteristics and its movements corresponding to the road conditions can be communicated to the neighboring vehicles by means of VANET.

3.1 Introduction

The main issues to be resolved in providing protection to pedestrians is the proper detection of the presence of objects crossing the road in front of vehicles. The objects to be detected could be automobiles, persons, or any obstacles. The distance of these objects from the vehicle can be measured using sensors (e.g., ultrasonic sensors) and the corresponding data should be processed and sent to the controller to take necessary action. High-resolution cameras are the most popular sensors used for the identification of pedestrians, and provide edge and contour information to the assistance system in automobiles. In order to get a clear vision of the environment, either visible or infrared cameras are preferred.

It is somewhat difficult to categorize the pedestrian protection system proposed in recent years into specific modalities: In some cases, there is no clear information, in other cases information is combined into a single module. So, in order to provide an organized and far-reaching review of the current system, we initially describe a conventional architecture of six distinct modules. Every module has its own objectives and duties in the framework, so by fixing every algorithm in the writing in its relating module, it will be simpler to think about and break down the recommendations. The modules that are available as follows:

1) Pre-processing
2) Candidate generation
3) Verification and fine-tuning
4) Classification
5) Tracking
6) Application

- *Pre-processing*: This stage takes input information and sets it up to advanced processing. The information mostly comes from a camera; however, in some instances, data is also procured from vehicle sensors, odometers, and so on. The assignments did are assorted, a few models or sensors synchronization, altering camera presentation time, gain, and adjustment.
- *Candidate generation*: This involves extricating areas of interest (pedestrians) from the picture, which is then to be sent to the classification module by excluding as many persons as possible on foot can be expected under the circumstances.
- *Verification and fine-tuning*: This process confirms and refines the persons classified as pedestrians, mentioned as identifications. The confirmation channels false positives utilizing criteria not covered in the classifier, while the refinement plays out a fine sub-division of the persons on foot (in reality, no outline arranged) so as to provide a precise separation estimation or to help the accompanying module tracking.
- *Classification*: The classification module generates a list of persons to be classified as either pedestrians or non-pedestrians.
- *Tracking*: This stage follows the identified people on foot along with time for few purposes, for example, avoiding fake or false discoveries, predicting the following walker position and bearing, and other significant tasks, for example, gathering the behavior of pedestrians.
- *Application*: This takes significance-level decisions (e.g., applying breaks and guiding) by utilizing the data provided by the past modules. This module speaks to a total territory of research, which incorporates driver observing as well as psychological issues, human–machine connection, and so forth (Figure 3.1).

Obviously, the point varies from general human identification frameworks, for example, surveillance applications or human–machine interfaces, for which a few disentanglements can be executed. For instance, the use of a static camera permits the utilization of foundation subtraction systems. The principal examinations on person-on-foot recognition for pedestrian protection systems (PPSs) were introduced in the late 1990s. From that point forward, PPSs have become a hot mechanical test that is of significant interest to governments, car organizations, providers, colleges, and research focuses. Accordingly, numerous papers tending to on-board person-on-foot recognition have been distributed, a couple of which in part review the cutting edge [1, 2, 3, 4].

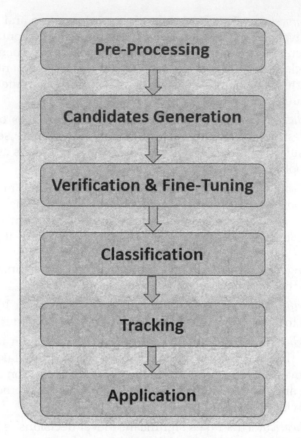

FIGURE 3.1
Workflow of the pedestrian protection system.

For example, in 2001, Gavrila [5] diagrammed a couple of existing frame-works around then, concentrating on the utilized sensors. In 2006, Gandhi and Trivedi [6] followed a similar methodology, however centered around the parts of impact forecast and passer-by conduct examination. Similar creators as of late introduced a study that surveys foundation advancements, sensors, and person-on-foot recognition approaches in a general vehicle safety setting, as opposed to concentrate particularly which is ready to discover. In contrast to different fields, for example, face or vehicle recognition, in which inside and outside surveys break down the calculations and effective frameworks have been introduced, walker discovery for Automatic Driver Assistance System (ADAS) does not have a thorough audit [7].

In an effort to diminish the amount of auto collisions or, if nothing else, restricting their effect, vehicles are progressively being fitted with dynamic security frameworks. Such frameworks only prove to be successful once the mishap is a ways into its event stage. To avoid this kind of circumstance from emerging in any case, it becomes important to foresee the related dangers and act as required. This stage requires great discernment combined with an intense comprehension of the driving condition. The proposed workflow structure offered an ascend to present uncommon sensors. (e.g., cameras, laser, and radar) into specific vehicles. These sensors have been intended to work inside a wide scope of circumstances and conditions (e.g., climate, glow, and so forth) with an endorsed set of variety limits [8]. Adequately identifying when a given working edge has been outperformed comprises a key parameter in the production of driving help frameworks that meet required unwavering quality levels. Considering this specific circumstance, a climatic perceivability estimation framework might be fit for measuring the most widely recognized working scope of installed exteroceptive sensors. This data is then used to adjust sensor activities to alarm the driver that the installed help framework is out of commission [9, 10, 11, 12] (Figure 3.2).

The coordination of correspondence innovation in best-in-class vehicles has started years prior: cars, telephones, and internet are increasingly dependent on latest advancements, just as Bluetooth connectors joining mobile phones are famous models. Be that as it may, the immediate correspondence between vehicles utilizing an ad hoc network, alluded to as infrastructure-vehicle correspondence (IVC) or specially appointed systems (e.g., vehicular ad hoc networks, known as VANETs), is a moderately new methodology. Contrasted with a cell framework, IVC has three key favorable circumstances: Lower inertness because of direct correspondence, more extensive inclusion, and having no administration expense. As of late, the guarantees of remote correspondences to help vehicular safety applications have prompted research around the world: The Vehicle Safety Communications Consortium building up to the DSRC innovation (USA), the Interest ITS Consortium (Japan), the PREVENT venture (Europe), or the 'System on Wheels' task (Germany).

FIGURE 3.2
A scenario of pedestrians crossing the road.

The particular properties of VANETs permits us to improve on alluring new administrations. Some current models in the two most significant zones – safety and solace – are as per the following.

1) *Comfort applications*: This sort of use improves traveller solace and traffic productivity or potentially advances the course to a goal. Models for this class are: Traffic data framework, climate data, corner store or eatery area and spot data, and intelligent correspondence, for example, internet access or music download.

2) *Safety applications*: Uses of this classification increment the security of travellers by trading safety important data by means of IVC. The data is either introduced to the driver or used to enact an actuator of a functioning security framework.

Model uses of this class are: Crisis cautioning framework, path evolving associate, crossing point coordination, traffic sign/signal infringement winding down, and street condition cautioning. The usage of this class generally requests direct vehicle-to-vehicle correspondence because of the stringent deferral prerequisites. Right now, it centers around systems administration issues which ought to be tended to for message trading between vehicles in VANETs. Since VANETs are a new subject of enthusiasm for logical and industry network, we unequivocally accept a complete overview learning about the theme is required. In the past, the creators had a survey of works in different convention stack layers, however, we will focus on the instruments rather than convention stack layers, and thereafter depict every system which can be actualized in various layers.

3.2 Literature Review

Providing an economical safety measure to pedestrians and also to the transport vehicles on the roads is a critical issue. VANET is a novel technique which is aiming to form an innovative automobile communication system to receive an alert about an accident. VANET is a specialized protocol which consists of a variety of wireless techniques (e.g., Bluetooth, ZigBee, WiFi, WiMax, etc.) to provide communication among the vehicles to the surrounding vehicles or infrastructure. Sharma [13] depicts that a GPS receiver can precisely position the information of a travelling vehicle. According to the GPS positioning values, the separation between vehicles can be measured using relative speed and position. The critical issue while considering these assumptions is that the corners of the vehicle may not be visible.

Deng [14] proposed a technique to follow multi-hopping in VANET's to support in computing the offload of vehicles. Considering the practical situation, a highly reliable multi-hop routing path is formed depending on the correlation theory of VANET. In this proposal, the task vehicle can maximize the utilization of VANETs to offload computation data through a multi-hop connection even if it has not entered the communication range of the mobile edge computing server. This can work well, but lags in time as it undergoes multiple hopping techniques.

Bouassida et al. [15] introduced a technique which control congestion by controlling the weight of the message transmission channel in wireless communication. The proposed approach minimized congestion and delay in critical message transmission. In this approach, priority is given to the usefulness and authenticity of messages and the velocity of vehicles. Shucong et al. [16] presented a cluster management scheme for heterogeneous VANET long-term evolution (LTE) networks. They analyzed the impact of VANET transmission in the heterogeneous network. A Markovian model was used to evaluate the vehicle-to-vehicle (V2V) traffic and compare it with standard methods. The results showed that V2V traffic and LTE traffic was lower compared with the standard schemes. The V2V traffic is efficiently served by the LTE network, and neither one affects the performance of the other, however, the network overhead was increased while the traffic was reduced.

Raut [17] proposes that a V2V-based methodology appears to be more reasonable for implementing an intelligent transportation system (ITS) but it will take more time to transmit safety messages related to traffic congestion to other coverage area which increases the delay in transmission of critical messages. The availability of vehicular to roadside unit (RSU) communication is beneficial for the transmission of traffic congestion messages to other coverage areas. In vehicle to RSU communication, if there is any sort of traffic congestion then vehicles who first detect the congestion will broadcast the message in their area so that all following vehicles will take alternative routes which help to control traffic congestion. The RSU of the congested area will send such critical messages to other RSUs in other coverage areas. The vehicles in other coverage areas will receive this critical message broadcasted by their RSU, and these vehicles will take alternative paths which helps to control traffic congestion. Using the concept of heterogeneous network in VANET, we can convey critical messages to other coverage areas with minimum delay as compared to V2V communication in VANET. One can use cellular network as it has a larger coverage area. [18]

3.3　Vehicular Ad Hoc Network

The systems that interconnect vehicles on the street are called vehicular ad hoc networks (VANETs). "A mobile ad hoc network (MANET) comprises versatile hubs that associate themselves in as decentralized, self-arranging way and may likewise set up multi-bounce courses. In the event that portable hubs are vehicles, this is called vehicular specially appointed system." "The principle focus of research in VANETs is the enhancements of vehicle safety by methods for inter vehicular communication (IVC)." A few distinct applications are rising in VANETs. These applications incorporate security applications to make driving a lot more secure. Portable trade and other data benefits will advise drivers about blockage, driving risks, mishaps, road turnings and parking lots. VANETs have a few unique differences when compared with MANETs; the hubs move with high speed on account of the topology changing quickly. VANETs are additionally inclined to a few unique attacks. Thus, the security of VANETs is essential. VANETs present numerous difficulties on innovation, conventions, and security, which increases the requirement of information regarding present scenario [19] (Figure 3.3).

A VANET is an innovation utilized by moving autos as hubs in a system to make a portable system. Every time VANET transfers the data to the vehicle of interest into a remote switch or hub, permitting autos roughly 100 to 300 meters from one another to interface and, thusly, make a system with a wide range. As autos drop out of the sign range and drop out of the system, different vehicles can participate, associating vehicles to each other with the goal that a versatile system is made. The primary frameworks that will incorporate this innovation are police and fire vehicles that will communicate with one another for safety purposes. VANETs fall under the class of arranging remote ad hoc's. In vehicular specially appointed systems, the hub might be a vehicle or the street side units. They can speak with one another by permitting the remote association up to a specific range. IVCs, otherwise called vehicular specially appointed systems (VANETs), have become well known as of late. A VANET is a unique sort of MANET (a MANET is a type of

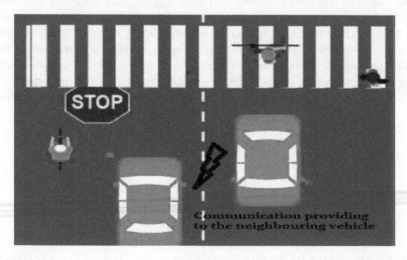

FIGURE 3.3
Communication among vehicles via VANET.

specially appointed remote system and is self-arranging system of versatile switches associated by remote connections) which uses vehicles as hubs. The principle contrast is that portable switches which fabricate the system are vehicles like autos or trucks. A few distinct applications are developing as to vehicular correspondences. For instance, safety applications for more secure driving, data administrations to inform drivers about the driving risks, and different business benefits in the region of the vehicle. Government, companies, and scholastic networks are taking a shot at empowering new applications for VANETs [20]. One of the primary objectives of VANET is to build street security by utilizing remote correspondences. To accomplish these objectives, vehicles go about as sensors and inform each other about strange and conceivably perilous conditions like mishaps, parking lots with turnings in road, and coatings. Vehicular systems intently take after impromptu systems in light of their quickly evolving topology. In this manner, VANETs require secure directing conventions. Various applications are remarkable to the vehicular setting. These applications incorporate security applications that will protect the drivers of portable trade through roadside benefits that can brilliantly advise drivers about traffic and organizations in the region. VANETs, particularly contrasted with MANETs, are identified by a few interesting angles. Hubs move with high speed, bringing about high paces of topology changes. On account of the quickly changing topology because of vehicle movement, the vehicular system utilizes a specially appointed system. The limitations and advancements are astoundingly extraordinary. From the system point of view, security and versatility are two huge difficulties. An impressive arrangement of misuses and assaults become apparent. Consequently, the security of vehicular systems is imperative. The developing significance of IVC has been perceived by the administration, partnerships, and scholarly networks. Government and industry participation has financed huge IVC organizations or activities, for example, Advanced Driver Assistance Systems and CarTALK 2000 in Europe, and FleetNet in Germany. VANETs present numerous difficulties on innovation, conventions, and security which increases the requirement for inquire about right now.

VANETs are relied upon to help an enormous range of versatile conveyed applications that go from traffic analyzing map and dynamic course intending to setting mindful promotion and record sharing. Considering the huge number of hubs that partake in these systems and their high versatility, discussions still exist about the plausibility of uses that utilizes start to finish multi bounce correspondence. The primary concern is whether the presentation of VANET directing conventions can fulfil the throughput and postpone necessities of such applications. Examinations of customary directing conventions for specially appointed portable systems (MANETs) exhibited that their presentation is poor in VANETs. The primary issue with these conventions, e.g., ad hoc on-request distance vector (AODV) and dynamic source routing (DSR), in which VANET should analyze the precariousness of the vehicle. The conventional hub-driven perspective on the courses (i.e., a set up course is a fixed progression of hubs between the source and the goal) prompting the broken courses within the sight of VANETs' high portability, as showed in Figure 3.4. Thus, numerous bundles are dropped, and the overhead because of course fixes or disappointment warnings altogether expands, prompting low conveyance proportions and high transmission delays.

One elective methodology is offered by topographical directing conventions, e.g., greedy–face–greedy (GFG), greedy other adaptive face routing (GOAFR), and greedy perimeter stateless routing (GPSR), which decouples sharing of data among the hub's. These conventions don't build up courses, yet utilize the situation of the goal and the situation of the neighbor hubs to advance information. Not at all like hub-driven steering, geological directing has a bit of leeway in that any hub that guarantees progress toward the goal can be utilized. For example, in Figure 3.4, land sending could utilize hub N2 rather than N1 to advance information to D. Notwithstanding the strength in better way, land sending does not perform well in city-based VANETs [21].

The concern is that, in many cases, the hub can't locate a next jump (i.e., a hub that is nearer to the goal than the present hub). The recuperation systems in the writing are regularly founded on planar diagram traversals, which were demonstrated to be insufficient in VANETs because of radio deterrents, high hub versatility, and the way that vehicle positions on streets as opposed to being consistently conveyed over a district.

VANET can offer different administrations and advantages to VANET clients and also maintaining excellent arrangement of hub's in the network. VANETs with interconnected vehicles and various administrations guarantee productive computerized foundations into many parts of our lives, from vehicle-to-vehicle, side of the road gadgets, base stations, traffic lights, etc. A

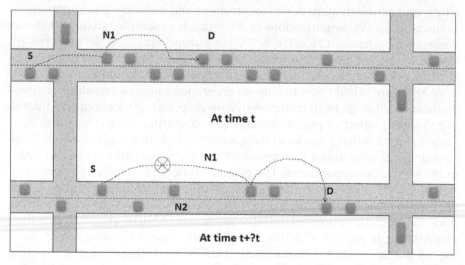

FIGURE 3.4
Frequent routing and their breaking in highly mobile VANETs.

system of many versatile and rapid vehicles through remote correspondence associations got better guidance electronically and in fact attainable and been created for stretching out customary traffic controls to fresh out of the box new traffic benefits that offer huge traffic-related applications. The data is secured and trade empowers life-basic applications, for example, the alarming usefulness during crossing point navigating and path consolidating, and along these lines assumes a key objective in VANET applications. The alluring benefits of VANET definitely will face higher risks if such systems don't consider preceding organization. For example, if safety-related messages are adjusted, disposed of, or postponed either deliberately or because of equipment failure, it could result in injury or loss of life.

Unlike traditionally wired networks which are protected by several lines of defence, such as firewalls and gateways, security attacks on such wireless networks may come from any direction and target all nodes. Therefore, VANETs are susceptible to intruders ranging from passive eavesdropping to active spamming, tampering, and interfering due to the absence of basic infrastructure and a centralized administration. Moreover, the main challenge facing VANETs is user privacy. Whenever vehicular nodes attempt to access some services from roadside infrastructure nodes, they want to maintain the necessary privacy without being tracked down for whoever they are, wherever they are, and whatever they are doing. It is considered as one of the important security requirements that requires more attention for secure VANET schemes, especially in privacy-vital environments. A number of security threats to VANETs have been addressed.

3.4 VANET Model Overview

There are numerous substances engaged with a VANET settlement and arrangement. In spite of the fact that by far most of VANET hubs are vehicles, there are different elements that perform fundamental activities in these systems. In addition, they can speak with one another from multiple points of view. Right now, a depiction about the most widely recognized elements such as the infrastructure and vehicle units in VANETs is provided in the subsequent section, where an examination is made of the distinctive VANET settings that can be found among vehicles and the rest of the elements. A few unique elements are normally accepted to exist in VANETs [22]. To comprehend the internal and related security issues of these systems, it is important to break down such elements and their connections. Figure 3.5 shows the ordinary VANET plot.

As shown in Figure 3.5, two different environments are generally considered in VANETs: Infrastructure environments and ad hoc environments.

Every time the condition of framework should be updated in which, substances can be for all time interconnected. It is primarily formed by those

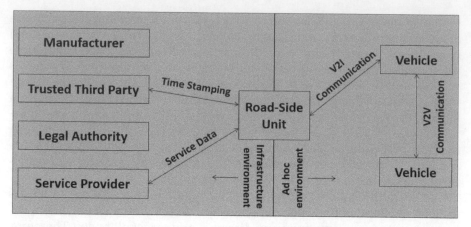

FIGURE 3.5
VANET model.

elements that deal with traffic or provide external assistance. On one hand, the manufacturers sometimes consider the inside architecture of VANET model. As a feature of the assembling procedure, they extraordinarily distinguish every vehicle. Then again, the legitimate authority is generally present in VANET models. Regardless of the various guidelines in every nation, it is constantly identified with two principle undertakings – vehicle enlistment and offense detailing. Each vehicle in a regulatory area is required to get enlisted once manufactured. Because of this procedure, the vehicle's position provides a tag. Then again, it additionally forms traffic reports and fines. Trusted third parties (TTP) are additionally present right now. They offer various administrative services like quality in execution or time stepping. The two makers and the authority are identified with TTPs in light of the fact that they, in the end, need their administrations (for instance, for giving electronic accreditations like passwords). Service providers are additionally considered in VANETs. They offer administration services that can be provided through the VANET. Location-based services (LBS) and digital video broadcasting (DVB) are two instances of such administrations for some specified conditions whose sporadic correspondences are set up from vehicles. From the VANET perspective, they are furnished with three unique gadgets. Right off the bat, they are outfitted with a correspondence unit (on-board unit) that empowers V2V and vehicle-to-infrastructure (V2I) interchanges. Then again, they have a lot of sensors to gauge their own status (e.g., fuel utilization) and its condition (e.g., elusive street, security separation). This sensorial information can be imparted to different vehicles to expand their mindfulness and improve street security. At long last, a trusted platform module (TPM) is regularly mounted on vehicles. These gadgets are particularly intriguing for security purposes as they offer dependable stockpiling and calculation. They, as a rule, have a solid inside clock and should be alter safe or if nothing else alter obvious (Raya et al., 2005, 2006) [23]. Right now, data (for example client certifications or pre-crash data) can be dependably put away.

As referenced previously, VANETs can interchangeably organize with any other kind necessity. Vehicles move at a generally fast speed and, then again, the high measure of vehicles present in a street could prompt a huge system. In this manner, a particular correspondence standard, called dedicated short-range communications (DSRC) has been created to manage such prerequisites (Armstrong Consulting Inc.). This standard determines that there will be a few specialized gadgets found on the side of the roads, called RSUs. Right now, VANETs become portals between the framework and vehicles, and vice versa.

3.5 VANET Settings

A few applications are empowered by VANETs, mainly affecting road safety. Inside this usage type, messages exchanged over VANETs are diverse in nature and reason. Considering this, four diverse correspondence examples can be observed, as discussed below.

V2V warning propagation (Figure 3.6): There are circumstances in which it is important to make an impression on a particular vehicle or a group of vehicles. For instance, when a mishap is identified, an admonition message should be sent to specific vehicles to expand traffic safety. Then again, if an emergency vehicle is coming, a message should be sent to the corresponding

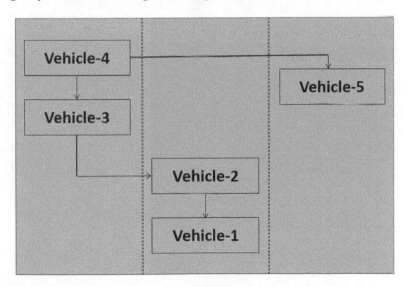

FIGURE 3.6
Propagation of warning.

vehicles before taking a move from vehicles. Right now, it would be simpler for the emergency vehicle to have an expressway. In the two cases, a directing convention is then expected to advance that message to the destination.

V2V group communication (Figure 3.7): In this example, only vehicles that have a few importance can partake in the correspondence. These highlights can be static (e.g., vehicles of a similar endeavor) or dynamic (e.g., vehicles on a similar territory in an interim period).

V2V beaconing (Figure 3.8): Beacon messages are sent occasionally to nearby vehicles. They contain the present speed, heading, braking use, and so forth of the sender vehicle. These messages are helpful to build neighbor mindfulness. Reference points are only sent to the vehicles which conveys the message in 1-bounce/step (e.g., they are not sent). Truth be told, they are useful for steering conventions, as they permit vehicles to find the best neighbor to route a message.

I2V/V2I warning (Figure 3.9): These messages are sent either by the framework (through RSUs) or a vehicle when a potential threat is identified. They are valuable for improving road safety. For instance, a notice could be sent by the framework to vehicles drawing closer to a crossing point when a potential crash may occur. There exist other correspondence designs over VANETs (e.g., identified with sight and sound access, area-based administrations, and so forth). Specifically, vehicles could utilize distinctive correspondence media like cell systems (for example GSM/GPRS) to get such administrations. Be that as it may, we will concentrate on V2V and V2I road safety correspondence designs over VANETs, as they will require more testing from a security perspective. Every correspondence design has an alternate arrangement of security prerequisites.

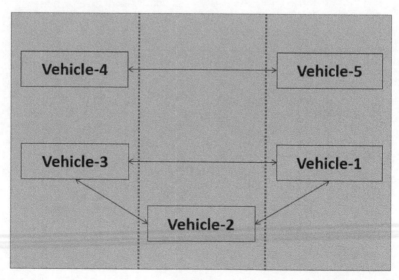

FIGURE 3.7
V2V group communication.

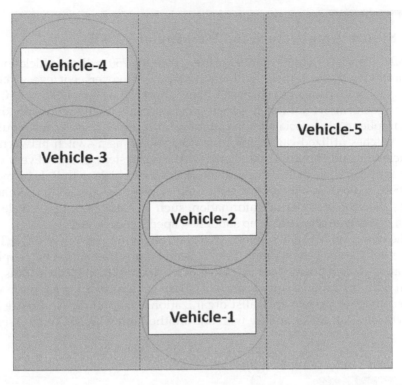

FIGURE 3.8
V2V channelling.

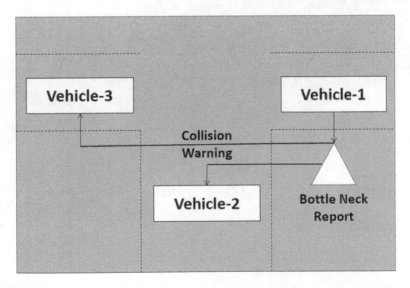

FIGURE 3.9
V2I threatening.

3.6 System Architecture and Working of VANETs

VANETs consist of a huge number of hubs, roughly 750 million vehicles on the planet today. These vehicles will require a power to oversee it, every vehicle can speak with different vehicles utilizing short radio signs DSRC (5.9 GHz), for range can arrive at 1 km, this correspondence is an ad hoc correspondence that implies each associated hub can move unreservedly, no wires required, the switches utilized called RSU, the RSU fills in as a switch between the vehicles out and about and associated with other system gadgets. Every vehicle has and on-board unit; this unit interfaces the vehicle with RSU by means of DSRC radios. Another gadget is the tamper-proof device (TPD); this gadget holds the vehicle's internal information, such as data about the vehicle like keys, driver characteristics, trip subtleties, speed, course, etc.

The design of VANET suggests that the conveying hubs in a VANET are either vehicles or base stations. Vehicles can be private (owned by people or privately-owned businesses) or open (open transportation implies transport, open administrations, squad cars, etc.). Base stations can have a place with government or private specialist organizations. As outlined in Figure 3.10, the vehicles can communicate with one another and with RSU conversely.

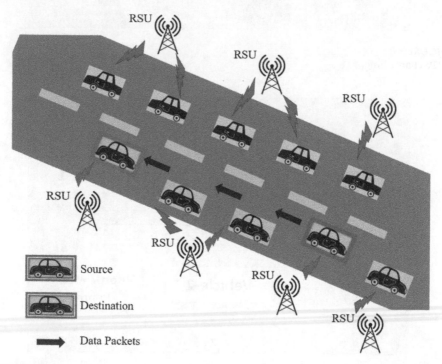

FIGURE 3.10
Architecture of VANET.

The size of VANETs is another element that separates them from one another. With a huge number of hubs disseminated all over, VANETs are probably going to be the biggest specially appointed portable systems. VANETs are a significant innovation for future improvements of vehicular correspondence frameworks. Such systems made out of moving vehicles, are equipped for giving correspondence among nearby vehicles and the roadside frameworks. Present day vehicles are furnished with figuring gadgets, occasion information recorders, receiving wires, and GPS, making VANETs feasible. VANETs can be utilized to help different functionalities, for example, vehicular safety, traffic congestion decreases, office-on wheels, and on-street notice. Most hubs in a VANET are portable, but since vehicles are commonly compelled to roadways, they have an unmistakable controlled versatility design. Vehicles trade data with their neighbors and steering conventions are utilized to proliferate data to different vehicles.

3.7 Application

VANET offers a few advantages to associations of any size. While such a system poses certain safety issues (for instance, one can't safely type an email while driving), this doesn't constrain VANET's potential as a profitable device. GPS and route frameworks can profit, as they can be incorporated with traffic reports to give the quickest course to work. A PC can transform congested driving conditions into productive work time by, for example, having work emails downloaded and read aloud by the on-board PC. It would likewise take into account free, voice over internet protocol solutions, for example, Google Talk or Skype between representatives, bringing down media communications costs.

3.8 Conclusion

The opportunities that a VANET present are unlimited. The future implementation of vehicular networks offers a tremendous opportunity to increase the safety of the transportation system and reduce road fatalities. The security-providing protocols find their scope in the current VANET applications to keep away from malicious activities that disrupt the performance of the network. Speed control, accident prevention, vehicle tracking, and police patrolling are some of the main application areas for the protocols proposed in this thesis. The scope of this research work not only pertains to the vehicular technologies but can be amended for applications in other network technologies, like mobile ad hoc networks and mobile wireless sensor networks, where mobility is present and security is a requirement.

References

1. Y. Abramson, Y. Freund: "Semi-automatic visual earning (SEVILLE): A tutorial on active learning for visual object recognition". In: *IEEE Conference on Computer Vision and Pattern Recognition*, San Diego, CA, USA (2005).
2. A. Broggi, M. Bertozzi, A. Fascioli, M. Sechi: "Shape-based pedestrian detection". In: *IEEE Intelligent Vehicles Symposium*, Dearborn, MI, USA (2000).
3. D. Gerónimo, A. López, A. Sappa, T. Graf: "Survey of pedestrian detection for advanced driver assistance systems". *IEEE Transactions on Pattern Analysis and, Machine Intelligence* 32(7), 1239–1258 (2010).
4. M. Enzweiler,M. Hummel, D. Pfeiffer, U. Franke: "Efficient stixel-based object recognition". In: *IEEE Intelligent Vehicles Symposium*, Alcalá de Henares, Spain (2012).
5. D.M. Gavrila, "Sensor-based pedestrian protection". *IEEE Intelligent Systems* 16(6), 2001.
6. T. Gandhi, M. M. Trivedi, "Vehicle Surround Capture: Survey of Techniques and a Novel Omni-Video-Based Approach for Dynamic Panoramic Surround Maps". *IEEE Transactions on Intelligent Transportation Systems* 7(3), SEPTEMBER 2006.
7. L. Itti, C. Koch, E. Niebur: "A model of saliency-based visual attention for rapid scene analysis". *IEEE Transactions on Pattern Analysis and, Machine Intelligence* 20(11), 1254–1259 (1998).
8. V. Cavallo, M. Colomb, J. Dor: "Distance perception of vehicle rear lights in fog". *Human Factors* 43, 442–451 (2001).
9. N. Dalal, "Finding people in images and videos", PhD thesis, Inst. Nat'l Polytechnique de Grenoble/INRIA Rho^ne-Alpes (2006).
10. B. Fardi, U. Schuenert, G. Wanielik: "Shape and motion-based pedestrian detection in infrared images: A multi sensor approach". In: *Proceedings of the IEEE Intelligent Vehicles Symposium*, 18–23 (2005).
11. B. Leibe, N. Cornelis, K. Cornelis, L. VanGool: "Dynamic 3D scene analysis from a moving vehicle". In: *Proceedings of the IEEE Conference on Computer Vision and Pattern Recognition*, 1–8 (2007).
12. M. Bertozzi, A. Broggi, A. Fascioli, A. Tibaldi, R. Chapuis, F. Chausse: "Pedestrian localization and tracking system with Kalman filtering". In: *Proceedings of IEEE Intelligent Vehicles Symposium*, 584–589 (2004).
13. M.S.P. Sharma, R.S. Tomar: "VANET based communication model for transportation system". In: *Symposium on Colossol Data Analysis and Networking (CDAN)*, IEEE (2016).
14. Z. Deng, Z. Cai, M. Liang: "A multi-Hop VANETs-assisted offloading strategy in vehicular mobile edge computing, special section on information centric wireless networking with edge computing for 5G and IoT". *IEEE Access*, 8 (2020).

15. M.S. Bouassida, M. Shawky: "On the congestion control within VANET". In: *1st IFIP Wireless Days*, 2008 1st IFIP Wireless Days, Dubai, 2008, pp. 1–5, doi: 10.1109/WD.2008.4812915.
16. S. Shucong Jia, X. Gu: "Analyzing and relieving the impact of FCD traffic in LTE-VANET heterogeneous network". 2014 *21st International Conference on Telecommunications (ICT)*, Lisbon, 2014, pp. 88–92, doi: 10.1109/ICT.2014.6845086.
17. Ch. M. Raut, S. R. Devane: "Intelligent transportation system for smartcity using VANET". In: *International Conference on Communication and Signal Processing*, April 6–8, IEEE (2017). DOI: 10.1109/ICCSP.2017.8286659.
18. P. Mutalik, et al.: "A comparative study on AODV, DSR, and DSDV routing protocols for intelligent transportation system (ITS) in metro cities for road traffic safety using VANET route traffic analysis (VRTA)". 2016 *IEEE International Conference on Advances in Electronics, Communication and Computer Technology (ICAECCT)*, Pune, 2016, pp. 383–386, doi: 10.1109/ICAECCT.2016.7942618.
19. J. Tarel, D. Aubert, F. Guichard: "Tracking occluded lane markings for lateral vehicle guidance". In: *IEEE CSCC'99* (1999).
20. L. Wischhof, A. Ebner, H. Rohling: "Information dissemination in self-organizing intervehicle networks". *IEEE Transactions on Intelligent Transportation Systems*, **6**(1), (2005).
21. M. M. Artimy, W. Robertson, XV. J. Phillips: "Connectivity with static transmission range in vehicular ad hoc networks". In: *3rd Annual Communication Networks and Services Research Conference (CNSR)* (2005).
22. J. Zhao, O. Cao: "VADD-vehicle-assisted data delivery in vehicular ad hoc networks". *IEEE Transactions on Vehicular Technology* **57**(3), 1910–1922, May 2008, doi: 10.1109/TVT.2007.901869.
23. M. Raya, J.-P, Hubaux: "The Security of Vehicular Ad Hoc Networks". In: *Proceedings of the 3rd ACM Workshop on Security of Ad Hoc and Sensor Networks* (2005), 11–21, https://doi.org/10.1145/1102219.1102223.

Part 2

Designing and Evaluation

4

Design of Platoon Controller with V2V Communication for Highways

Anuj Abraham, Nagacharan Teja Tangirala, Pranjal Vyas,
Apratim Choudhury, and Justin Dauwels

CONTENTS

ABSTRACT   The capability of vehicles to communicate wirelessly provides an opportunity for innovations in the transport sector. Platooning of vehicles is one of the widely studied topics in the research community. Vehicle platooning aims to increase the capacity of roads, improve traffic safety, and reduce fuel consumption. The goal of this chapter is to introduce the reader to the subject of vehicle dynamics on modelling and control aspects and observe the behavior with vehicle-to-vehicle (V2V) communication constraints. The proposed algorithms are presented in general form and are applicable to any vehicle.

Most of the work on platooning investigates the control of vehicles in pairs based on the concept of look-ahead platooning. This chapter emphasizes the design of proportional integral derivative (PID) controllers and model predictive control (MPC) based on a combination of constant distance (CD) and constant headway time (CHT) policies to operate a heavy-duty vehicle platoon. In addition to basic cooperative adaptive cruise control (CACC), the controller is tested and verified for carrying out splitting and merging maneuvers.

Furthermore, the controller performance is analyzed with V2V communication constraints between the vehicles. Computer simulations were carried out to test the effectiveness of the controller. MATLAB, VISSIM, and a network simulator (NS3) are combined to form an integrated simulation. The platooning operations and merging and splitting maneuvers are replicated as realistically as possible while considering the effects of neighboring V2V equipped traffic. The speed, acceleration, and inter-vehicular distance profiles of the platoon were observed. It is evident from the profiles that the followers track the header vehicle and maintain a constant distance and time gap in the presence of frequent speed variations by the header vehicle.

The packet delivery ratio (PDR) of the platoon vehicles is determined in the case of external disturbance due to the V2V-equipped surrounding vehicles. The PDR can be used to establish a correlation between communication and controller performance. Finally, this chapter also presents an analysis of packet drops in vehicle platooning using V2V communication, where deliberate communication failures are introduced through NS3.

4.1 Introduction

Potential researchers in the transportation sector and academia, who are working in the vehicle-to-everything (V2X) domain, will be interested to see the impact of V2X on real-life traffic conditions. The primary motive of this chapter is to present innovations on connected transportation systems that demonstrate a new way of achieving improvements in safety, fuel consumption, and mobility.

Advancements in V2X technology involve V2X application development to enhance traffic systems. However, experimentations are not practical in the prototype development stages. To counter this problem, the research community has proposed several simulators that enable the study of V2X applications. Since there is no standalone package to study V2X applications, several integrated architectures are designed using existing software packages. In this research study, an integrated simulator framework is set up for testing V2X protocols and applications. Several V2X-based applications, such as green light optimized speed advisory (GLOSA), platooning, and intersection collision avoidance, are studied using the simulator. This enables us to study the impact of V2X on real-life traffic conditions. This chapter focuses on the design of the platoon controller with vehicle-to-vehicle (V2V) communication for highways.

Truck platooning is one of the more complex V2X-based applications and involves vehicles traveling together in a pattern. In long-distance travel, platooning provides a significant reduction in fuel consumption, thereby reducing the cost of operation and air pollution. The most common pattern is a straight line with vehicles moving one behind the other in the same lane on highways. The first vehicle sets the pace, and the remaining vehicles adjust their speed to match the first vehicle. A significant proportion of air-drag is reduced by making the vehicles follow each other in close-range [1]. This results in a reduction in fuel consumption. Throughout the world, freight transport consumes a million tons of fuel every year. Hence, even a modest decrease in fuel consumption makes a significant difference on a broader scale. Additionally, platooning increases roadway capacity [2].

The challenge in platooning is to develop a capable controller that can keep the vehicles at a close-range, track the abrupt speed changes of the first vehicle, and avoid collisions among the platoon members [3]. Most of the earlier work on the platoon controller design entails the use and development of an adaptive cruise control (ACC) strategy [4]. In the ACC system, a controlled vehicle obtains the speed and position of the immediate neighboring vehicles to calculate its control input. Radar communication is used for communication between the platoon members. Due to the inherent time-delays in radar communication, it is unreliable in emergencies [5]. To overcome the short-comings of the ACC strategy, researchers proposed an advanced version of ACC, called cooperative adaptive cruise control (CACC). CACC strategy uses wireless communication, such as 5.9 GHz dedicated short-range communications (DSRC) or 5G to exchange data among the platoon members. The use of wireless communication allows low latency and high-frequency data exchange of a wide range of data such as GPS position, inertial measurement unit data, and the control actions like throttle or brake. Access to more data helps in implementing advanced control strategies. In this chapter, the development of a CACC strategy using a proportional integral derivative (PID) and model predictive control (MPC) is described.

The controller design is based on a combination of constant distance (CD) and constant headway time (CHT) policies. V2V and vehicle-to-infrastructure (V2I) communications are collectively termed as V2X communication. There are two types of devices used in V2X communication: Vehicles are equipped with on-board units, and the infrastructure is equipped with road-side units. On-board units transmit the vehicle state information, and road-side units transmit data depending on the infrastructure at which they are installed (e.g., a road-side unit at an intersection transmits the traffic signal data). To ensure an acceptable quality of service in V2X communication, the wireless standard must support high-frequency and low-latency data exchange.

To demonstrate vehicular communication as realistically as possible, we have chosen NS3 for network simulations. It contains the model libraries for simulating wireless access for vehicular environments (WAVE) architecture. WAVE is a family of standards that allow secure V2X communication in a high-speed transportation environment. The platoon and surrounding traffic are modelled in VISSIM [6], with MATLAB being used to connect VISSIM and NS3. The simulation results indicate a stable performance of the controller in real-life traffic situations. The neighboring V2V equipped traffic effect is accounted to analyze the controller behavior. This will provide a good starting point for the automotive industry as well as academic researchers to build advanced controllers for truck platooning.

4.2 Concept of Look-Ahead Platooning

In look-ahead platooning, the vehicles are controlled with information from the first vehicle and the vehicle which is immediately ahead of it. This approach provides better control and reduces the risk of collision. The clearance between the platoon members is generally implemented using two policies, i.e., the constant distance policy and constant time gap policy. In the constant distance policy, the gap between the platoon members is a constant value decided at the design stage of the controller. In the constant time gap policy, the gap is proportional to the speed of the platoon. This study tries to combine both CD and CHT policies to utilize the best of both.

Each platoon vehicle is addressed with a specific name depending on the position of the vehicle. Figure 4.1(a) shows a platoon of three heavy-duty vehicles connected with V2V communication. The vehicle in the front is the *header* of the platoon. A certified driver drives the header heavy-duty vehicle. The destination, speed and the lane are at the driver's discretion. Irrespective of the length of the platoon, the first heavy-duty vehicle is always called as a header. The rest of the heavy-duty vehicles are called followers. They follow the header and are not allowed to make any decisions like joining or leaving the platoon without the permission of the header. The arrows indicate the information exchange between the header and all the followers which are required to implement the ACC strategy.

In Figure 4.1(b), a four heavy-duty vehicle platoon is shown. The header of the platoon acts as a leader for follower 1, follower 1 acts as a leader for follower 2, and follower 2 acts as a leader for follower 3. As there is no vehicle after follower 3, it does not act as a leader to any vehicle. In the CACC strategy, every follower uses double feedback to determine the control action. The first feedback is obtained from the header, and the second feedback is obtained from the leader. Here, the arrows indicate the additional feedback provided by the leaders to their next followers. It must be noted that the leader vehicle only provides the feedback; hence it is indicated with a single-headed arrow originating from the leader to the follower. The data exchange arrows are indicated from the controller design perspective. Through V2V communication, all vehicles exchange data with all the other vehicles. The

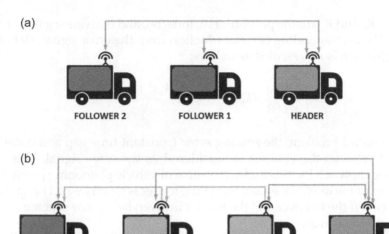

FIGURE 4.1
(a) A three heavy-duty vehicle platoon indicating the header and followers. (b) Leader and header of a follower vehicle in a four heavy-duty vehicle platoon.

terminology described in Figure 4.1(b) is extendable to any length of the platoon.

4.3 Modelling of the Platoon

The mathematical modeling for a platoon of vehicles is explained in this section. The platoon is considered to be analogous to a mass-spring-damper system [7], which can be represented as follows:

$$m_n\ddot{x}_n + b_n\dot{x}_n = u_n. \tag{1}$$

where index n denotes nth vehicle of the platoon, m_n is the mass of the vehicle, b_n is the damping constant, x_n is the longitudinal position, and u_n is the input information.

To implement a platoon controller with CD and CHT policies, this paper utilizes the PID controller, which can be formulated as follows:

$$u(t) = k_p e(t) + k_i \int_0^t e(t)dt + k_d \frac{d}{dt} e(t). \tag{2}$$

where K_p, K_i, and K_d are the proportional, integral, and derivative gains of the system. The corresponding transfer function from the error signal $e(t)$ to the input signal $u(t)$ is represented as follows:

$$G(s) = \frac{U(s)}{E(s)} = k_p + \frac{k_i}{s} + k_d s = \frac{k_d s^2 + k_p s + k_i}{s}. \qquad (3)$$

For this control problem, the spacing error (constant time gap and constant distance) between the vehicles is considered as the error signal. This is a pragmatic approach for longitudinal control of vehicle platooning [7–10]. The spacing error between the follower and the leader is given by ε_{fl}. The spacing error between the follower and the header is given by ε_{fh}. The spacing errors are calculated as follows:

$$\varepsilon_{fl} = x_l - x_f - hd_{fl} - (hw_{fl} \cdot v_f) = dx_{fl} - hd_{fl} - (hw_{fl} \cdot v_f)$$
$$\varepsilon_{fh} = x_h - x_f - hd_{fh} - (hw_{fh} \cdot v_f) = dx_{fh} - hd_{fh} - (hw_{fh} \cdot v_f). \qquad (4)$$

where x_l, x_f, and x_h, respectively, represent the GPS positions of the header, leader, and follower; hd_{fl} and hw_{fl} represent the desired constant distance and headway time between the leader and the follower, respectively. Similarly, hd_{fh} and hw_{fh} represent the desired constant distance and headway time between the header and the follower, respectively.

For a set of n vehicles, the constant distance and headway time between the header and the follower can be calculated as follows:

$$hd_{fh} = (n-1) \times hd_{fl}$$
$$hw_{fh} = (n-1) \times hw_{fl}. \qquad (5)$$

Using the concept mentioned in [7], the force required for moving a platoon of vehicles can be written as follows:

$$m\dot{u} = F_x - mg\sin\theta - f_r mg\cos\theta - \frac{1}{2}C_{air}(u + u_w)^2. \qquad (6)$$

where $C_{air} = \rho A_f C_d$ is a constant, F_x is the tractive force of the vehicles in the platoon, C_d is the drag co-efficient of the vehicle, A_f is the frontal area of the vehicle, u_w is the wind velocity, θ is the angle of inclination, u is the vehicle forward velocity, m is the mass of the vehicle, and g is the acceleration due to gravity. u_w is the wind velocity, which is positive for a headwind and negative for a tailwind. The drag co-efficient for the vehicle ranges from about 0.2 to 1.5.

4.4 Proposed Controller Methodology

A feasible method of PID control and MPC strategy for truck platooning is proposed. The main features involved in the formulation of the controller are as follows:

- The controller ensures that the follower vehicles track the speed and maintain the desired gap between the vehicles.
- At each time instant, every vehicle broadcasts its speed and position information.
- Each follower computes its own control input using the information from the header and its preceding vehicle.
- In an ideal communication scenario, the follower vehicle will always have the information of the header and its preceding vehicle for every time instant.
- With V2X communication constraints, there will be some packet drops in the farthest vehicles of the platoon. The farthest vehicles take more time to match the setpoint speed due to the packet drops. This will lead to undershoots and overshoots in the inter-vehicular distances.

4.4.1 PID Controller Design

Using the variables described, the acceleration of the vehicle with the PID control law can be formulated in as,

$$a_f = \left[\frac{(a_h + a_l)k_d + (v_h - v_f)k_p + (v_l - v_f)k_p + k_i(x_h - x_f - hd_{fh} - ((hw_{fh} \cdot v_f)) + k_i(x_l - x_f - hd_{fl} - ((hw_{fl} \cdot v_f))}{(C_{air} \cdot u^0 + 2k_d)} \right]$$

where, (7)

a_h = Acceleration of the header vehicle	v_h = Velocity of the header vehicle	C_{air} = Air drag of the car
a_l = Acceleration of the preceding vehicle	v_l = Velocity of the preceding vehicle	m = Mass of the vehicle
a_f = Acceleration of the follower vehicle	v_f = Velocity of the follower vehicle	u^0 = Vehicle forward velocity

hd_{fh} = Distance between header and follower vehicles

hd_{fl} = Distance between preceding and follower vehicles

hw_{fh} = Headway time between the header and follower vehicles

hw_{fl} = Headway time between the preceding and follower vehicles

The corresponding state-space formulation is given by:

$$
\begin{bmatrix} \dot{dx}_{fh} \\ \dot{dx}_{fl} \\ v_f \\ a_f \end{bmatrix} = \begin{bmatrix} 0 & 0 & -1 & 0 \\ 0 & 0 & -1 & 0 \\ 0 & 0 & 0 & 1 \\ k_i/m & k_i/m & -2k_p/m & (-2k_d - C_{air}.v_f)/m \end{bmatrix} \begin{bmatrix} dx_{fh} \\ dx_{fl} \\ v_f \\ a_f \end{bmatrix}
$$

$$
+ \begin{bmatrix} 1 & 0 & 0 & 0 \\ 0 & 0 & 1 & 0 \\ 0 & 0 & 0 & 0 \\ k_p/m & k_d/m & k_p/m & k_d/m \end{bmatrix} \begin{bmatrix} v_h \\ a_h \\ v_l \\ a_l \end{bmatrix} \tag{8}
$$

$$
+ \begin{bmatrix} 0 & 0 \\ 0 & 0 \\ 0 & 0 \\ -k_i/m & -k_i/m \end{bmatrix} \begin{bmatrix} hd_{fh} + hw_{fh}.v_f \\ hd_{fl} + hw_{fl}.v_f \end{bmatrix}.
$$

The distances dx_{fl} and dx_{fh} are calculated as given in Eq. 9:

$$
dx_{fh} = x_h - x_f, \ dx_{fl} = x_l - x_f. \tag{9}
$$

Using the above information along with v_f and x_f, the output vector $y = \begin{bmatrix} dx_{fh} & dx_{fl} & v_f & a_f \end{bmatrix}^T$ can be represented as:

$$
\begin{bmatrix} dx_{fh} \\ dx_{fl} \\ v_f \\ a_f \end{bmatrix} = \begin{bmatrix} 1 & 0 & 0 & 0 \\ 0 & 1 & 0 & 0 \\ 0 & 0 & 1 & 0 \\ 0 & 0 & 0 & 1 \end{bmatrix} \begin{bmatrix} dx_{fh} \\ dx_{fl} \\ v_f \\ a_f \end{bmatrix} + \begin{bmatrix} 0 & 0 & 0 & 0 \\ 0 & 0 & 0 & 0 \\ 0 & 0 & 0 & 0 \\ 0 & 0 & 0 & 0 \end{bmatrix} \begin{bmatrix} v_h \\ a_h \\ v_l \\ a_l \end{bmatrix}. \tag{10}
$$

4.4.2 MPC Control Strategy

The MPC controller aims to maintain inter-vehicular distance and headway time between the vehicles for two cases, namely unconstrained and con-strained optimization problems. The controller starts adjusting the control signal ahead of reference changes. In a constrained optimization problem, the controller allows the flexibility to restrict the process variables, mainly the acceleration and speed of each vehicle to user-defined limits. The advantage of MPC over PID controller design is that it can handle consecutive packet losses and delays in the communication between the vehicles.

In this section, a kinematic model [11] for the design of an MPC controller is given by:

$$x_{k+1}^i = x_k^i + v_k^i T_s + \frac{1}{2} a_k^i T_s^2 \tag{11}$$

$$v_{k+1}^i = v_k^i + a_k^i T_s$$

where T_s is the sampling time.

The objective of the platoon vehicles is to track the desired speed of the header and maintain inter-vehicular distances based on the combination of CD and CHT topologies [12]. The state vector is formed with the information of three vehicles only. That is, the position and velocity of the header, the $(j-1)$th vehicle, and jth vehicle along with the acceleration of the header. The input is the acceleration of $(j-1)$th vehicle and jth vehicle. Now, considering an n-vehicles platoon, no extension is needed for the state vector or the input vector. The discrete-time model is formed as:

$$X_{k+1} = AX_k + Bu_{k-1} + B\Delta u_k \tag{12}$$

$$Y_k = \begin{bmatrix} x_1 - x_{j-1} \\ x_{j-1} - x_j \\ v_1 - v_{j-1} \\ v_{j-1} - v_j \end{bmatrix} = CX_k \tag{13}$$

where:

$$A = \begin{bmatrix} 1 & T_s & 0 & 0 & 0 & 0 & \frac{1}{2}T_s^2 \\ 0 & 1 & 0 & 0 & 0 & 0 & T_s \\ 0 & 0 & 1 & T_s & 0 & 0 & 0 \\ 0 & 0 & 0 & 1 & 0 & 0 & 0 \\ 0 & 0 & 0 & 0 & 1 & T_s & 0 \\ 0 & 0 & 0 & 0 & 0 & 1 & 0 \\ 0 & 0 & 0 & 0 & 0 & 0 & 1 \end{bmatrix}$$

$$B = \begin{bmatrix} 0 & 0 \\ 0 & 0 \\ \frac{1}{2}T_s^2 & 0 \\ T_s & 0 \\ 0 & \frac{1}{2}T_s^2 \\ 0 & T_s \\ 0 & 0 \end{bmatrix} \quad C = \begin{bmatrix} 1 & 0 & -1 & 0 & 0 & 0 & 0 \\ 0 & 0 & 1 & 0 & -1 & 0 & 0 \\ 0 & 1 & 0 & -1 & 0 & 0 & 0 \\ 0 & 0 & 0 & 1 & 0 & -1 & 0 \end{bmatrix}$$

$$X = \begin{bmatrix} x_1 & v_1 & x_{j-1} & v_{j-1} & x_j & v_j & a_1 \end{bmatrix}^T \tag{14}$$

$$u = \begin{bmatrix} a_{j-1} & a_j \end{bmatrix}^T ; u_k = u_{k-1} + \Delta u_k \tag{15}$$

Using the formulated discrete time model, the predictions are made for N steps. N is called the prediction horizon. Here, only one step control is considered, that is $\Delta u_{k+n} = 0$ (n>0). The prediction for N steps is given by:

$$\hat{Y} = \begin{bmatrix} Y_{k+1} \\ Y_{k+2} \\ \cdot \\ \cdot \\ \cdot \\ Y_{k+N} \end{bmatrix} = \Phi X_k + \Gamma u_{k-1} + \Gamma \Delta u_k \tag{16}$$

where:

$$\Phi = \begin{bmatrix} CA \\ CA^2 \\ \cdot \\ \cdot \\ \cdot \\ CA^N \end{bmatrix} \Gamma = \begin{bmatrix} CB \\ CB + CAB \\ \cdot \\ \cdot \\ \cdot \\ \sum_{i=0}^{N} CA^i B \end{bmatrix}$$

The reference R can be generated based on different topologies. For CD and CHT topologies used in the design, R is given by:

$$R = \begin{bmatrix} (hd + v_{j-1}T)(j-2) & (hd + v_j T) & 00 & \cdots \\ \cdots & (hd + v_{j-1}T)(j-2) & (hd + v_{j-1}T) & 0 & 0 \end{bmatrix} \tag{17}$$

According to the control goal in [13], a cost function is formed as:

$$J = (\hat{Y} - R)^T (\hat{Y} - R) + \lambda \Delta u^T \Delta u \tag{18}$$

where λ is the tuning parameter of the chosen value 0.1.

The optimal solution is obtained by minimizing the cost function J. However, it should be noted that only the desired acceleration a_j is applied to the jth vehicle, and a_{j-1} is discarded.

4.5 Platoon Maneuvers

The platoon controller obtained in the previous section can be used for platoon operation. However, the formation and dissolution of the platoon are also essential. The basic platoon maneuvers are splitting, merging, and lane change operations [14]. Splitting divides a platoon into two successive ones in the same lane and is initiated by the follower vehicles. Merging combines two successive platoons in the same lane and is initiated by the headers of two platoons. Any other maneuver, such as a vehicle leaving or joining the platoon at an arbitrary position, can be realized by executing a series of merging and splitting protocols in an appropriate manner.

Every maneuver requires a coordinated exchange of data among the vehicles through V2X communication. There are two ways of implementing communication. Some of the earlier studies have recommended the use of V2I communication. However, it is inefficient to install road-side units throughout the highway. Additionally, they pose safety concerns due to communication drops in tricky situations, such as hilly regions or tunnels. Another way of communication is with V2V communication using on-board units. Merging and splitting maneuvers based on V2V communication are described in this chapter, as shown in Figure 4.2.

There are three phases, the *appeal* phase, the *reply* phase, and the *implementation* phase (ARI). A vehicle makes an appeal to the header and provides the information relating to its request. It can be either a split or merge request. If there is a communication failure, the appeal is sent again in the next time instant. If the reply is affirmative, the vehicle is instructed to carry out the maneuver. If rejected, the vehicle can choose to either repeat the appeal or wait for a certain time, depending on the situation. Sample protocols for

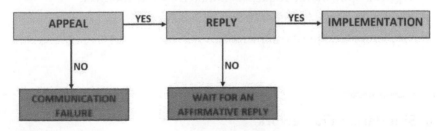

FIGURE 4.2
The ARI protocol for platoon maneuvers.

FIGURE 4.3
(a) Merge protocol of the platoons. (b) Split protocol of the platoons.

merge and split maneuvers are shown in Figure 4.3(a) and Figure 4.3(b), respectively.

- **Appeal**: Split/merge request is initiated by any follower.
- **Reply**: Approval/rejection from the header.
- **Implement**: Request completed.

4.6 Simulation Framework

In order to study the effect of V2X technology on various traffic scenarios, there does not exist any self-supporting simulation tool. Fortunately, it is possible to integrate a combination of simulation software. One of the widely

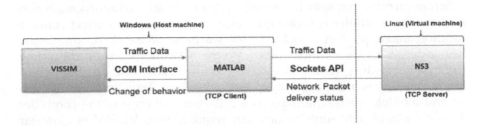

FIGURE 4.4
Block diagram of the simulator environment.

used open-source traffic simulators is the simulation of urban mobility (SUMO) [5]. However, it does not support left-hand-driving traffic simulations, which are essential to test applications for Singapore road networks. Hence, this chapter uses one of the most versatile microscopic simulators, namely VISSIM, for the simulation of traffic in Singapore. VISSIM offers a resolution as small as 50 ms and supports real-time data exchange when coupled with other simulators. To simulate V2X communication between the vehicles, NS3 is used. MATLAB helps in coordinating and synchronizing the simulations between VISSIM and NS3.

Figure 4.4 illustrates the block diagram of the simulator integration used in this work [15]. MATLAB and VISSIM communicate with each other using VISSIM's component object model (COM) interface. Both software options are installed on Windows OS, but NS3 is developed for Linux. Hence, a Linux virtual machine is set up on Windows for running NS3. The transmission control protocol/internet protocol (TCP/IP) socket application programming interface (API) is used to establish information exchange between NS3 and MATLAB. The controller application is coded in MATLAB. The position and velocity of platoon vehicles are obtained at every simulation time step from VISSIM using the COM interface. Using this obtained data, MATLAB computes control action in terms of speed. At the same time, MATLAB sends the vehicle data to NS3 to perform communication simulations and gets the results from NS3. The results include the information regarding the packet delivery status from the header to the followers. Using this information, MATLAB provides speed input to the vehicles in VISSIM through COM only when the packet status is successful. More details regarding the simulation functionality are described in [16].

4.7 Key Observations

All the platoon applications are tested on roadways chosen from the Singapore traffic network. The layouts are modelled in VISSIM with appropriate traffic volume data obtained from the Land Transport Authority (LTA), Singapore.

Before carrying out tests in the presence of realistic communication constraints, the controller is thoroughly evaluated, assuming perfect communication among platoon vehicles. In the warm-up phase of the simulation, heavy-duty vehicles are introduced into the network. After a certain time, the first heavy-duty vehicle is assigned as the header and all the remaining heavy-duty vehicles combine to form a platoon. Traveling through traffic and multiple intersections poses a different challenge to the controller. The positions of heavy-duty vehicles are acquired from VISSIM as Cartesian coordinates with respect to a local reference frame, and the inter-vehicular distance is computed as Euclidean distance.

1) Once the controller performance is evaluated with an ideal communication setup, communication constraints are introduced. Every heavy-duty vehicle of the platoon is assigned to a node in NS3 for the purpose of network simulations. Since the simulations only involve platoon vehicles, the NS3 channel scheduler is programmed to allow the control channel to have continuous access. When surrounding vehicles are considered, all the vehicles are assumed to communicate in the control channel with surrounding vehicles broadcasting randomly in the control channel to add interference. The on-board unit parameters are set according to the DSRC WAVE protocol [16 and 17]. The transmitter power level is set to 23 dBm, and the message data rate is set to 6 Mbps. In addition to platoon vehicles, surrounding vehicles are programmed to broadcast packets at random time instants to place additional load on the channel, in order to establish the impact of constant communication interference on the platoon controller. This situation would be realistic when V2X technology becomes mainstream.

2) The packet delivery ratio (PDR) is a good indicator of network performance. For a platoon, PDR is defined as the ratio of the number of packets broadcasted by the header to the number of packets received successfully by the follower. PDR is calculated for an individual follower, and the graph is generated by combining the PDR of vehicles that are within a certain distance from the header. It has been observed that the heavy-duty vehicles that are far from the header take more time to converge toward the desired inter-vehicular distance.

3) After settling to the desired setpoint, there is a stable behavior exhibited by the platoon heavy-duty vehicles. From the observations, it is safe to conclude that the controllers can keep the platoon safe in a real-world traffic scenario with a real-world communication setup. The concept of a safe gap is very important in the platooning of vehicles as they serve as a backup protocol to prevent vehicles from colliding with each other if frequent packet drops are observed.

4) In the existing CACC controller [5], the vehicles maintain a GAP_{SAFE} between them, which is determined by the speed and the maximum deceleration ability of the individual vehicles. Given that the packet drops are unavoidable in V2V communication, the risk of collision increases by manifold. The exchange of information for switching between modes takes more time than the systems that work on a single controller logic. The proposed controller described in Eq. 7 uses both CD and CHT policies. Hence, the control applications work seamlessly under any speeds and further decreasing the chances of collisions.

5) Furthermore, when packet drops are introduced into the platoon communication to study the controller performance, it has been observed that a predictive algorithm can handle consecutive packet losses and delays in communication. Also, the MPC controller maintains constant inter-vehicular distance and headway time between the vehicles for the defined optimization constraints.

The platoon vehicle is given a setpoint speed to accelerate and decelerate within some time intervals. Then, communication failures are deliberately introduced in any one vehicle of the platoon. When the platoon accelerates, no collisions were observed even when the duration of the failure was increased to 4 s. Whereas when the platoon decelerates, collision is observed with 1 s of communication failure [18].

4.8 Challenges

1) One of the major challenges in platoon coordination is string stability. The header vehicle of the platoon aims to maintain constant inter-vehicular distance between the followers. Normally, radar detection is used for the measurement of inter-vehicular distances between the platoon members. For any sudden variation in transient or steady states (when accelerated or decelerated), it is difficult to ensure the string stability of platoon members. String instability can also take place in the presence of communication delays, consistent packet drops, and network interference.

2) It is also a challenge to maintain string stability when a platoon vehicle is carrying out splitting and merging maneuvers. Proper investigation is necessary while performing the platoon maneuvers to study the effect of the intended acceleration or deceleration of the vehicle.

An enhanced formulation is required for the use of Euclidean distance to account for the curvature of roads. This will help in studying large-scale platooning for various complex urban networks throughout the world.

The curvature of roads should matter for accurate maintenance of inter-vehicular gap. One of the major challenges in an urban environment would be to maintain a platoon while it is traveling through intersections as traffic lights may turn red when only half the platoon has crossed. Maybe there needs to be a platoon priority at the intersection and more scope for investigation.

3) The controller design becomes complex when extended to support a variety of vehicles for heterogeneous platooning.

4) It is difficult to model a lower-level controller considering the engine torque effect on platooning.

4.9 Summary

The concept of platooning is one of the important research topics within the V2X domain due to the benefits it brings to road transport. This work addresses the platooning problem by proposing a look-ahead platooning method using the PID controller and MPC strategy. The controller is designed with CHT and CD policies combined to use the best of both the policies. It is observed that the controller performs well in all the scenarios and causes no collisions among the platoon members at steady and transient states and maintains both constant distance and time gap. Moreover, an MPC strategy ensures systematic handling of constraints and provides significant performance improvements over the PID controller when successive packet drops occur. The possibility of collision during communication failure depends on the speed change magnitude. This analysis helps to determine the additional sensors required to provide fail-safe alternatives in case of communication failures. Additionally, ARI protocols for platoon splitting and merging are proposed to facilitate all the platoon formation and dissolution. It is analyzed that the ARI protocols are effective in carrying out platoon maneuvers reliably.

Acknowledgments

This work has been supported by the NTU-NXP Intelligent Transport System Test-Bed Living Lab fund S15-1105-RF-LF from the Economic Development

Board (EDB), Singapore. The research has been done in the School of Electrical and Electronic Engineering at Nanyang Technological University, Singapore.

References

1. A. Alam, J. Martensson, K.H. Johansson, Experimental evaluation of decentralized cooperative cruise control for heavy-duty vehicle platooning, *Control Engineering Practice* 38 (2015) 11–25.
2. A. Alam, B. Besselink, V. Turri, J. Martensson, K.H Johansson, Heavy-duty vehicle platooning for sustainable freight transportation: A cooperative method to enhance safety and efficiency, *IEEE Control Syst.* 35 (2015) 34–56. doi:10.1109/MCS.2015.2471046.
3. D. Swaroop, J.K. Hedrick, Constant spacing strategies for platooning in automated highway systems, *J. Dyn. Syst. Meas. Control.* 121 (1999) 462. doi:10.1115/1.2802497.
4. I.A. Ntousakis, I.K. Nikolos, M. Papageorgiou, On microscopic modelling of adaptive cruise control systems, *Transp. Res. Procedia.* 6 (2015) 111–127. doi:10.1016/j.trpro.2015.03.010.
5. M. Amoozadeh, H. Deng, C.N. Chuah, H.M. Zhang, D. Ghosal, Platoon management with cooperative adaptive cruise control enabled by VANET, *Veh. Commun.* 2 (2015) 110–123. doi:10.1016/j.vehcom.2015.03.004.
6. M. Fellendorf, VISSIM: A microscopic simulation tool to evaluate actuated signal control including bus priority, in *64th Inst. Transp. Eng. Annu. Meet.*, Dallas, TX, Oct. 1994.
7. Z. Gacovski, S. Deskovski, Different control algorithms for a platoon of autonomous vehicles, *IAES-Int. J. Robot. Autom.* 3 (2014) 151–160. doi:10.11591/ijra.v3i3.5591.
8. D. Swaroop, J.K. Hedrick, C.C. Chien, P. Ioannou, A comparison of spacing and headway control laws for automatically controlled vehicles, *Veh. Syst. Dyn.* 23 (1994) 597–625. doi:10.1080/00423119408969077.
9. R. Rajamani, H.S. Tan, B.K. Law, W. Bin Zhang, Demonstration of integrated longitudinal and lateral control for the operation of automated vehicles in platoons, *IEEE Trans. Control Syst. Technol.* 8 (2000) 695–708. doi:10.1109/87.852914.
10. Y.A. Harfouch, S. Yuan, S. Baldi, An adaptive approach to cooperative longitudinal platooning of heterogeneous vehicles with communication losses, *IFAC-PapersOnLine.* 50 (2017) 1352–1357. doi:10.1016/j.ifacol.2017.08.225.
11. Galip Ulsoy, Huei Peng, Melih Cakmakc, *Automotive Control Systems*, Cambridge University Press, Cambridge, 2012.
12. Shengling Shi, Mircea Lazar, On distributed model predictive control for vehicle platooning with a recursive feasibility guarantee, *IFAC-PapersOnLine* 50(1) (2017) 7193–7198.
13. D. Honc, R. Sharma, A. Abraham, F. Dusek and N. Pappa, Teaching and practicing model predictive control, *IFAC-PapersOnline* 49(6), (2016) 34–39.
14. S. Dasgupta, V. Raghuraman, A. Choudhury, T.N. Teja, J. Dauwels, Merging and splitting maneuver of platoons by means of a novel PID controller, in *2017 IEEE Symp. Ser. Comput. Intell.*, IEEE, Hawaii, USA, 2017, pp. 1–8. doi:10.1109/SSCI.2017.8280871.

15. A. Choudhury, T. Maszczyk, C.B. Math, H. Li, J. Dauwels, An integrated simulation environment for testing V2X protocols and applications, *Procedia Comput. Sci.* 80 (2016) 2042–2052. doi: 10.1016/j.procs.2016.05.524
16. A. Choudhury, T. Maszczyk, M.T. Asif, N. Mitrovic, C.B. Math, H. Li, J. Dauwels, An integrated V2X simulator with applications in vehicle platooning, in *IEEE Conf. Intell. Transp. Syst. Proc., ITSC*, Rio de Janeiro, Brazil, 2016, pp. 1017–1022. doi:10.1109/ITSC.2016.7795680.
17. J.B. Kenney, Dedicated short-range communications (DSRC) standards in the United States, *Proc. IEEE.* 99(7) July (2011) 1162–1182. doi:10.1109/JPROC.2011.2132790.
18. N.T. Tangirala, A. Abraham, A. Choudhury, P. Vyas, R. Zhang, J. Dauwels, Analysis of packet drops and channel crowding in vehicle platooning using V2X communication, in *2018 IEEE Symposium Series on Computational Intelligence (SSCI)*, Bangalore, India, (2018). 281–286.

5

Using Computer Simulations for Quantifying Impact of Infrastructure Changes for Autonomous Vehicles

Priyanka Mehta, Pranjal Vyas, Anuj Abraham, Shyam Sundar Rampalli, Usman Muhammad, Shashwat, and Justin Dauwels

CONTENTS

ABSTRACT Interest in autonomous vehicles (AVs) has geared up globally in recent years with a focus on safety and low fuel consumption. Several industries and research organizations are collaborating on the development of AVs. Apart from this, government transportation agencies around the world are involved in bringing AVs on roads by performing various physical and virtual tests. An essential aspect of deploying AVs is increasing passengers' safety. It is highly advisable to validate the AVs with the current road infrastructure and propose efficient AV-friendly road infrastructure, including several lanes, lane widths, and bus bay designs. This chapter considers the bus bay design structure specific to Singapore for validation and proposing new bus bay designs. Two new bus bay designs are proposed and compared with existing bus bay designs. The comparison is performed based on queue length, bus arrival, and bus exit time. We have implemented the simulation using an integrated simulator developed in-house by combining a virtual test drive (VTD) and a robot operating system (ROS). We deployed real traffic data from the Ang Mo Kio (AMK) district area in Singapore for this purpose. The results show that the queue length developed in the proposed bus bay designs are shorter than the existing bus bay design, especially in high traffic density scenarios.

5.1 Introduction

Autonomous vehicles (AVs) are a highly anticipated innovation in the intelligent transportation industry, and we see a significant surge in its growth. Scientists, engineers, and policy-makers around the world are working together to reduce the gap where AV becomes a reality to overcome urban mobility challenges. Urban mobility and associated congestion is a recognized, yet increasingly complex, challenge. The transportation and infrastructure industries are collaborating and defining their investment priorities to adapt to the upcoming advancements of AVs. We are also seeing AVs as a forthcoming option for improving the first-mile and last-mile connectivity. The micro transportation model, such as shared e-scooters and e-bikes, has received mixed reviews in terms of safety and comfort from both users and non-users of the service. As all these involved industries race to make AVs a reality, we will witness significant infrastructure changes in urban cities adapting to a driverless future. The technological advances in a virtual simulation, internet of things (IoT), artificial intelligence (AI), and machine learning linked with city transportation has created myriad opportunities for cities to overcome these challenges [1]. In this chapter, we present computer simulations for quantifying the impact of infrastructure change for using AVs in an urban city, in a virtual environment. The chapter also discusses different designs of infrastructure and their impact on traffic flow where AV is a part of that traffic. We have investigated the effect of infrastructure changes in an urban city, with emphasis on crowded locations like bus stops near train stations.

The rest of this chapter is organized as follows. In Section 5.2, we discuss the literature related to our study. In Section 5.3, we describe the simulation environment where we discuss various software and the method we used for simulation. In Section 5.4, we propose different bus bay designs, which also contain the proposed designs and the method adopted to simulate them. This section also provides the assumptions made for simulations. In Section 5.5, we discuss the results where we analyze the performance of designs based on the performance indicator of queue length with different parameters. Finally, in Section 5.6, we provide a summary and ideas for future research.

5.2 Literature Review

Advances in the field of robotics, high performing computational capabilities, and the ability to analyze large sets of data are promising signs for a future with AVs as a mainstream reality. We also see AVs as an upcoming option for improving first-mile and last-mile connectivity [1]. On the other hand, there are significant advances in the multidisciplinary field of intelligent transportation systems (ITS) combined with information technology [2]. A large proportion of the existing work on intelligent transportation systems focuses on assisting human drivers in avoiding collisions at intersections [3, 4]. However, there are also studies with AVs as an upcoming option for improving first-mile and last-mile connectivity, which comes with the potential for safer travel and reduced congestion [5, 6]. As the involved industries race to make AVs a reality, we will witness significant infrastructure changes in urban cities.

Meyer et al. [7] explore various scenarios of potential autonomous mobility-based transportation concerning accessibility. The authors conclude with the possibility of a significant increase in accessibility from AVs. The research also mentions that depending on the magnitude of gained capacity from AVs, an equivalent of 15 years of infrastructure investment may be required. Hence, stakeholders must gain a realistic view of changes to be able to make informed decisions. This brings up an underlying question, are our cities ready for AVs? For example, Hobert et al. [8] discuss the use of vehicle-to-everything (V2X) technology with AVs and how the enhanced infrastructure and combination of both technologies (V2X and AV) will be required to increase safety and traffic efficiency. There is also research underway on AV path planning [9, 10] that focuses on the shortest path algorithm and path planning in unknown environments.

There have been studies on implementing future technologies in a simulation environment to test their viability and experiment with different test conditions.

Developing a city model has its roots in traditional city mapping [11], and combining digital data with spatial and temporal knowledge provides endless simulation possibilities. Wang et al. [12] have used the virtual environment to model the road surfaces to allow easier road distance calculation. Donikian [13] has developed a model of the virtual urban environment to study the realistic behavior of car drivers and pedestrians and their interaction with each other. Virtual models for transportation sectors have focused on different aspects, ranging from traditional transportation modeling to covering some particular topics like building textures [14] and lightings [15].

With regard to the impact of AVs on urban areas, studies show that AVs require less headway distance and less lane width than conventional vehicles because of their high precision in driving. In this way, AVs can enhance the capacity of road infrastructure [16], providing a better opportunity for green and public spaces. On the other hand, the high traffic density due to less headway distance and lane width may create problems for the safety and comfort of pedestrians and cyclists, especially in mixed traffic conditions [17]. A research article by Gavanas [18] concludes that urban changes are crucial for preparing for a future with AVs.

From the literature review, we found that there is research underway on quantifying different aspects of AV research and development. But there is a gap in the details of urban changes required for AVs. With this study, we attempt to fulfill the literature gap of specific infrastructure that could be adopted for integrating AVs into existing traffic environments.

In our study, we have performed simulations using a combination of simulation environment software. The AV simulation software provides the capability of deploying different sensors such as virtual light detection and ranging (LIDAR) with physics-based rendering, vision sensors, and inertial measurement units. These sensor measurements help the vehicles to interpret the surrounding environment used for localization, perception, path planning, and collision avoidance, etc. Traffic simulation enables populating vehicles based on real traffic conditions. The integration of traffic data into the simulation helps to test and analyze different scenarios to quantify the impact of AV in cases like J-walking, valet parking, and last-mile transportation. The simulation results help in expressing the challenges involved in adapting an urban city for AVs in its environment.

5.3 Simulation Environment

An integrated simulator developed in-house is used for analyzing the effect of infrastructure change. The simulator consists of two types of software: Virtual test drive (VTD) [19] and robot operating system (ROS) [20]. The simulator architecture is designed to provide a straightforward data exchange among the software and replicate real life through simulation. The architecture can link with external libraries for exchanging data and controlling different aspects of simulation in real time. VTD software is utilized for modeling vehicle dynamics. VTD facilitates the usage of static and dynamic environments and various sensor models, such as camera, LIDAR, and object sensors, for testing autonomous driving scenarios. VTD is interfaced with ROS, which controls the AV based on the various scenarios. Autoware [21], an open-source self-driving software based on ROS, is employed, which is used as the autonomous driving system for the AV models in the simulator. Figure 5.1 illustrates the architecture of an integrated simulator platform for AV deployment. The next section describes the details of VTD–ROS.

5.3.1 VTD–ROS Interface

This interface has been developed by making ROS nodes, which connect to VTD for receiving the information. VTD software provides the required components, such as traffic objects, road layout, and infrastructure, for scenario-based AV testing while using physics-based sensors. The components that need to be controlled at a low frequency, such as maneuverability of traffic vehicles and pedestrians, state of traffic lights, and generation of new traffic players, are handled by the simulation control protocol (SCP). High-frequency data exchange, such as traffic objects and sensor information between VTD and external libraries, is performed using runtime data bus (RDB). Both these interfaces have been embedded in ROS-based architecture to incorporate and further develop open-source libraries such as Autoware. Autoware is an ROS-based open source library which provides the complete pipeline for the AI of self-driving, from localization to path following, and has been validated in many studies. Autoware is modular, which makes it possible to plug in and plug out any specific AI algorithm to be tested in the whole pipeline by catering to the inputs and outputs of the other modules. The interfaces developed in ROS for interacting with VTD are the shared memory interface and network interface.

The shared memory interface is used for the exchange of high-frequency sensor data and is based on shared memory. It is implemented through RDB in VTD. RDB is a VTD proprietary binary communication protocol which exports simulation data in its own package format. Figure 5.2 illustrates the flow diagram of the shared memory interface with a camera and LIDAR sensor. ROS nodes are developed to extract both camera and LIDAR sensor data, which write their data in RDB format to two different memory

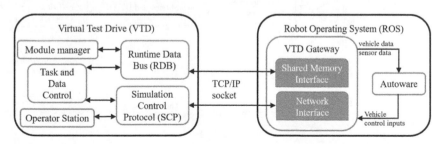

FIGURE 5.1
Simulation architecture using VTD and ROS.

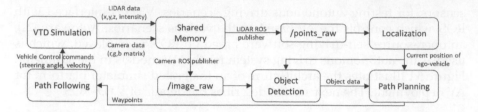

FIGURE 5.2
Data flow diagram of VTD–ROS interface.

segments. The RDB package for a sensor contains the simulation time, frame number, package ID, pixel format, message size, and data array. The sensor RDB package is then read through an ROS node that connects to the particular shared memory segment, parses the package, formats it into an ROS sensor message, and publishes it onto the respective ROS topic. This sensor data is read at each simulation step, and the frequency for LIDAR can be as high as 30 Hz.

The network interface simulation in VTD is externally controlled using a network interface built on proprietary SCP, which uses transmission control protocol/internet protocol. The American Standard Code for Information Interchange (ASCII) data in SCP commands uses XML syntax, which is easy to read and modify. Full control of the simulation is possible by solely using SCP, as VTD has a comprehensive suite of these commands for each aspect of the simulation. The synchronization of simulation time with the processing of algorithms for self-driving is achieved using SCP as well.

In our current test scenario, SCP sends the vehicle control commands of velocity and steering angle to the VTD driver model. Figure 5.3 shows the flow of simulation control of the integrated simulator, and Figure 5.4 illustrates the connection establishment between the VTD–ROS interfacing with LIDAR beam rendering. In the next section, we propose bus bay designs as an alternative for AV amiable infrastructure.

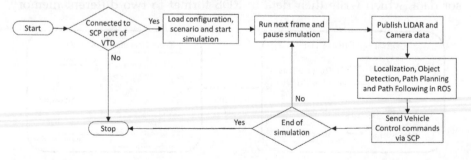

FIGURE 5.3
Flowchart for simulation control of the integrated simulator.

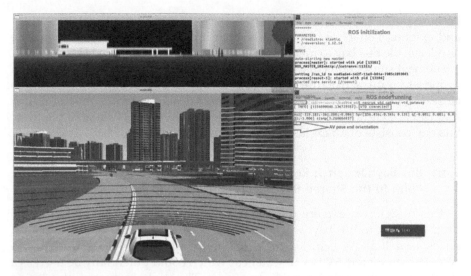

FIGURE 5.4
Snapshot illustrating VTD–ROS connection.

5.4 Proposed Bus Bay Designs

In this section, we will study the proposed infrastructure changes in the form of three bus bay designs (including baseline design) to accommodate AV as an integral part of public transportation. This study developed a model with road networks, bus bays, traffic signals, and buildings in the environment in Ang Mo Kio (AMK) district of Singapore. The chosen area of interest is a residential area containing major urban elements like rapid mass transits, bus interchanges, schools, parks, and a public library. Other than that, AMK has a large population and is the eighth-most populated district in Singapore, leading us to make a safe assumption that the area utilizes public transport considerably. All these factors made AMK a good sample for this study, which has proposed various public transport bus bay designs and AV shuttle buses to aid public transportation.

Bus bays are specially constructed areas that provide for the loading and unloading of passengers. They are separated from traffic lanes and are situated away from the normal sections of a roadway [22]. Bus bays allow through traffic to move freely without the obstruction of buses, and they should be provided primarily in high-traffic-volume or high-speed roadways such as urban expressways, or in heavily congested downtown areas where a large number of passengers board and alight. Bus bays can be found everywhere in Singapore, and they are also the prevailing public transport infrastructures in several major megacities in Asia, including Hong Kong, Beijing, and Tokyo, because public transit is a primary transport mode in these cities. These entry and exit areas enable a bus to safely enter a bus bay from the shoulder lane and leave the bus bay to merge into traffic on the shoulder lane [23].

The next subsections discuss the three proposed designs. The first design proposes an infrastructure that shares a bus bay and AV shuttle bus and is closest to the baseline condition. The second design is an elongated bus bay that can accommodate two buses and one AV shuttle bus. Finally, the third bus bay consists of an exclusive AV lane parallel to the bus bay.

5.4.1 Bus Bay Design 1: Redesign of Taxi Pick-up/Drop-off Point to the Shared Bay for Bus and AV Shuttle

In this design, we explore the possibility of redesigning the taxi pick-up/drop-off point to bus bay which shall also be shared by the public transport buses as well as AV vehicles, making it more adaptable for the future with AVs. The simulated AV is a 12 seater car which could potentially be used as a shuttle bus aiding existing public transport. The study focuses on the design parameters and its consequential effect on traffic. This design is referred to as "shared bus-AV bay" in the rest of the chapter. Figure 5.5 shows the location of the redesign.

The design of the bus bay plays a crucial role in traffic flow [24]. In the study area of interest, we realized that the exit outside the rapid mass transit has a taxi pick-up/drop-off point. We have redesigned this taxi point to a shared bus bay. The Land Transport Authority (LTA), Singapore, has designed specifications for bus bays and other road elements. The agency has set guidelines on the location and sizing of bus bays, which are factors for bus frequency and site constraints, if any. The width of a bus and the availability of bus shelters are essential parameters for the safety of passengers. But this study focuses on traffic flow where the whole design of the bus bay is of importance. Hence, we have designed the bus stop to be as similar as possible to the existing bus in an attempt to replicate the baseline.

FIGURE 5.5
Location of proposed shared AV bus bay.

As per the specifications outlined in Table 5.1, the sizing of the bus bay type depends on the frequency of the bus at the bay per peak hour. Figure 5.6 is a schematic representation of bus bay design, where X_1 and Z_1 are constant lengths, but Y_n varies depending on bus frequency. Figure 5.6 also contains a bird's eye view of a bus bay as captured on Google Earth.

These sizes are taken as reference for local microsimulation and should be modified when a similar study attempt is made for other cities.

To determine the appropriate bus bay size, we first collected the data on bus services available in the area of interest. There are 22 bus services which begin their journey from the AMK interchange. The frequency range at peak hours and non-peak hours for each bus is available from the bus service websites. We extracted this information for all the buses in the area to develop Table 5.2. We used this data to deduce an average minimum headway. In this

TABLE 5.1

Measurement Parameters of Bus Bay [25]

Frequency of buses per peak hour	Bus bay type	
1–39	One single bus bay	18 m (Y_1)
40–59	One double bus bay	32 m (Y_2)
60–79	One triple bus bay	46 m (Y_3)
80–99	Two double bus bays (with 10 m separation on the straight kerb)	32 m + 32 m +10 m
100 and above	One double bus bay plus one triple bus bay (with 10 m separation on the straight kerb)	18 m + 46 m + 10 m

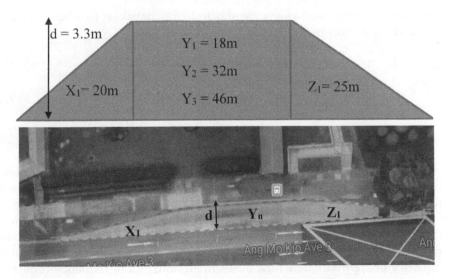

FIGURE 5.6
Schematic diagram of bus bay in Singapore with Google Earth image for reference.

TABLE 5.2

Minimum Average Headway and Turning Ratio

Bus service number	Minimum headway in minutes	Turning direction
130, 130A	8	Right
133	8	Right
135, 135 A	8	Straight
136	9	Right
138	7	Right
166	8	Right
169	2	Right
22	8	Straight
24	8	Straight
25	5	Left
261	3	Left
262	7	Straight
265	5	Right
269	5	Right
73, 73 A	8	Left
86	8	Right
	Average minimum headway ~7 min	Turning ratio: Straight – 5/22 Right – 10/22 Left – 7/22

case, the minimum headway is the shortest time between two buses of the same route (bus number). This can be represented as:

$$min_i = \min(t_{i1}, t_{i2}, t_{i3}, t_{in}) \tag{1}$$

$$min_i = \min(t_{j1}, t_{j2}, t_{j3}, t_{jn}) \tag{2}$$

$$\text{Average minimum headway} = \bar{X}(min_i, min_j, \dots min_n) \tag{3}$$

For AMK, we found the average minimum headway of approximately seven minutes. This implies that a passenger must wait for a minimum of seven minutes, on average, between buses of the same route. This waiting time is in agreement with the study [26], which assumed a headway of less than ten minutes. We used this information to feed into the model and create a bus line where buses frequented the bus stop every seven minutes, making it a frequency of approximately nine buses in an hour. Such frequency requires a single bus bay design, which is a total of 63 m in length and 3.3 m in width [27], as shown in Table 5.2 and Figure 5.6.

(a) Assumption 1 – AV frequency at the proposed bus bay: We proposed a taxi pick-up/drop-off point be redesigned as a shared AV – bus bay. Since AVs are not mainstream yet, we have made assumptions on the sizing of bay inspired by the guidelines from LTA's bus bay sizing recommendations. As seen, for the frequency of nine buses per peak hour, a single bay bus size is sufficient, but we are also simulating penetration of AVs in the same traffic. We have assumed the frequency of AVs to be 3.5 minutes to reduce the waiting time to half (seven minutes for buses).

(b) Assumption 2 – sizing of bay sharing bus and AV: We need extra space for the shuttle bus as compared to the current scene where the bay is used only for buses. Hence, we have assumed the bus bay length to be 70 m, which is the mean value of a single bus bay (63 m) and a double bus bay (77 m) (see Figure 5.6). Figure 5.7 shows the baseline design (Design 1). The design is a proposed redesign of taxi pick-up/drop-off to a shared bay for bus and AV shuttles and is closest to the baseline condition.

5.4.2 Bus Bay Design 2 – Elongated Bus Bay

Design 1 proposes a longer bus bay which can house two buses and one AV. The length of the bus bay is back-calculated in the model by first placing two virtual buses and one AV at the bay and then evaluating whether they can maneuver seamlessly. With this method, the length of the bus bay comes out to be equivalent to the length of a triple bus bay, i.e., 91 m. In this simulation,

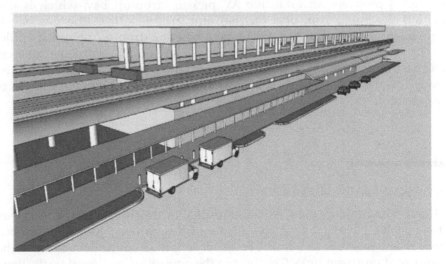

FIGURE 5.7
Design 1: Proposed redesign of taxi pick-up/drop-off point to the shared bay for bus and AV shuttle.

FIGURE 5.8
Design 2: Elongated bus bay.

only the design of the bus bay is different from the shared bus-AV bay. All the remaining parameters, like traffic volume, signal phase timings, and passenger volume, are maintained exactly the same as design 1. Figure 5.8 illustrates this design, and it is referred to as "elongated bus bay" in the rest of the chapter.

5.4.3 Bus Bay Design 3 – AV Pick-up/Drop-off Point Separated from Bus Bay

Design 3 proposes an exclusive AV pick-up/drop-off bay, which is just behind the public transport bus bay line. The length of the bus bay is equivalent to a single bus bay. The taxi pick-up/ drop-off point is parallel to the bus bay and shares the same length. Again, the traffic parameters and assumptions are maintained in this simulation as well. Figure 5.9 illustrates this design, and it is referred to as "exclusive AV point" in the rest of the chapter.

5.5 Results and Discussion

This section presents the simulation results for the three proposed bus bay designs. We have analyzed the impact of design over queue length, bus arrival, and exit time in a link. The queue length of an incident refers to the number of upstream links (i.e., links in the opposite direction of traffic flow) experiencing congestion [28].

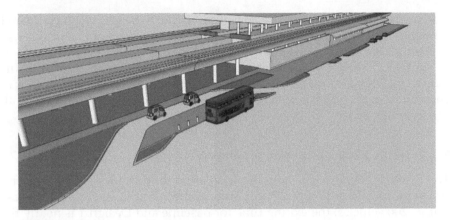

FIGURE 5.9
Design 3: Exclusive AV point–AV pick-up/drop-off point separated from bus bay.

Let B_t be the arrival of AV at the bus bay, E_t, be the time when the AV exits the link. Queue length is defined in terms of the average number of vehicles behind the AV. The impact of the introduction of AV in an urban environment for the three scenarios considered is analyzed in terms of B_t, link exit time E_t and queue length.

In this analysis, a road link is considered for computing the above parameters. Here, AV is made to start from the beginning of this link. The time parameters B_t and E_t are computed with and without traffic considerations. Queue length is computed in terms of an average number of vehicles behind AV at a regular interval of 30 seconds till E_t. The above parameters are calculated for different traffic radii around the AV, namely 1.5 km, 2.5 km, and 5 km, and different vehicle densities, namely 200, 500, and 800. Other traffic around the AV is generated randomly, and the data is extracted from multiple simulation runs. The speed and acceleration of AV are considered to be 20 km/h and 4 m/s². The AV stops at the bus bay for 8 seconds, and the length of the link is 444 m.

The following results illustrate the impact of different designs with various traffic densities:

5.5.1 Traffic Radius of 1.5 km around AV

In this scenario, we have a traffic radius of 1.5 km and a density of 200 vehicles within that radius. It is noted that the vehicles get queued up before 60 seconds where the AV has not reached the bus bay. Between 60 and 90 seconds, the AV reaches the bus bay with and without traffic. The changes in bus bay infrastructure dimension result in reduced queue length for the elongated bus bay design (Design 2) and the exclusive AV point design (Design 3) when compared with the shared bus-AV bay (Design 1).

Also, the parameters such as E_t and B_t are found to be less when compared with those of Design 1. This analysis is performed for different vehicle densities and traffic radiuses around the AV.

Figure 5.10 shows queue length at various traffic densities for a traffic radius of 1.5 km around the AV, and Table 5.3 shows the time the bus took to enter and exit the bay. It is seen that the bus took longer to enter and exit with higher traffic density in the baseline scenario.

5.5.2 Traffic Radius of 2.5 km around AV

The results in Figure 5.11 show the simulation for densities of 200, 500, and 800 vehicles in a radius of 2.5 km around the AV. It is seen that with less traffic (200 vehicles) the exiting time for baseline and Design 1 is nearly the same. The exclusive AV point design (Design 3) stands out with the least queue lengths for 60-sec and 90-sec time steps. The bus arrival and exiting time is better in the elongated bus bay design (Design 2) and the exclusive AV point design (Design 3) as compared to the shared bus-AV bay design (Design 1), which is closest to the baseline conditions. Table 5.5 illustrates the details of the results.

Table 5.4 shows that the designs perform similarly with lower traffic densities, but the performance of Design 3 is significantly better with scenarios of high vehicle density. Similar results are seen in Table 5.5.

5.5.3 Traffic Radius of 5 km around AV

The results in Figure 5.12 show the simulation for densities of 200, 500, and 800 vehicles in a radius of 5 km around the AV. With traffic radius of 5 km around the AV and density of 500 vehicles, the queue length for the exclusive AV point design (Design 3) was longer in 60 seconds (by one car) but went down significantly as compared to the shared bus-AV bay design (Design 1) in 90 seconds.

With regard to infrastructure changes, we see that most cities with developed public transportation systems create and use guidelines for placing amenities like bus stops and bus bays. The Transit Cooperative Research Program has published material that discusses the parameters to be considered for designing of bus stops and bus bays for US cities [29]. These parameters include factors like bus operations, traffic, as well as the safety and comfort of passengers. The research also discusses the locations of bus bays based on the speed and frequencies of vehicles. Previous researchers have developed models in VISSIM of AVs, especially for connected vehicles [30] [31]. There are studies with AV as a potential public transport aiding technology in context to Singapore [26]. The research focuses on strategic planning for future AV influx and understanding the effect of AV penetration in crowded areas like public transport bus bays

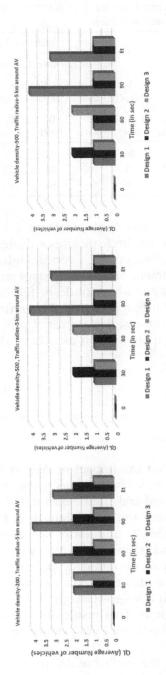

FIGURE 5.10

Queue length with different vehicle density and traffic radius of 1.5 km around AV.

TABLE 5.3

Bus Arrival and Exit Time from Bus Bay for traffic with a Radius of 1.5 km around Autonomous Vehicle

			With traffic					
			Traffic radius = 1.5 km around AV					
	Without traffic		Vehicle density – 200		Vehicle density – 500		Vehicle density – 800	
Scenarios	B_t (sec)	E_t (sec)	B_t (sec)	E_t (sec)	B_t (sec)	E_t (sec)	B_t (sec)	E_t (sec)
Design 1	66	80	84	106	92	120	120	148
Design 2	68	83	74	100	86	112	101	126
Design 3	70	85	76	96	80	103	112	130

right outside local train stations. Such locations see the highest influx of daily commuters, and the research proposes designs for aiding public transport with AV as an added option. Even though level 4 autonomy for AVs is still at the development and testing stage around the world, the research has proposed designs which could be feasible to implement considering the constraints of urban planning. This study uses AMK district data for simulation. Still, the insights from the project could be adapted for other districts in Singapore as well as other urban cities with some modifications.

5.6 Summary

In this chapter, we analyzed the impact of AVs in real traffic scenarios. As the road infrastructure profoundly impacts the ease of AV usage, three bus bay designs are proposed and simulated for different traffic densities. We compared the travel time and queue length of the proposed designs when AV is introduced in the city and shares space with public transportation infrastructure. A reduction in travel time by 5% is observed when AV has an exclusive lane for pick-up and drop-off in parallel with bus bays. A reduction of 27% in the number of vehicles queuing was also observed when the exclusive AV point design (Design 3) is compared to the shared bus-AV bay design (Design 1). This result made the performance of the exclusive AV point design better for both performance indicators analyzed in this study.

It should be noted that the exclusive AV point design requires more area, which can be a constraint in poor urban cities. It is an important

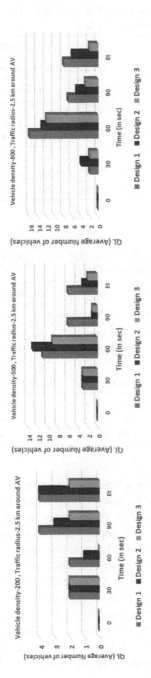

FIGURE 5.11

Queue length with different vehicle density and traffic radius of 2.5 km around AV.

TABLE 5.4

Bus Arrival and Exit Time from Bus Bay for Traffic with a Radius of 2.5 km around Autonomous Vehicle

	Without traffic		With traffic					
			Traffic radius = 2.5 km around AV					
			Vehicle density – 200		Vehicle density – 500		Vehicle density – 800	
Scenarios	B_t (sec)	E_t (sec)	B_t (sec)	E_t (sec)	B_t (sec)	E_t (sec)	B_t (sec)	E_t (sec)
Design 1	66	80	75	101	94	112	120	150
Design 2	68	83	73	95	80	106	112	137
Design 3	70	85	74	93	80	103	98	120

TABLE 5.5

Bus Arrival and Exit Time from Bus Bay for traffic with a Radius of 5 km around Autonomous Vehicle

	Without traffic		With traffic					
			Traffic radius = 5 km around AV					
			Vehicle density – 200		Vehicle density – 500		Vehicle density – 800	
Scenarios	B_t (sec)	E_t (sec)	B_t (sec)	E_t (sec)	B_t (sec)	E_t (sec)	B_t (sec)	E_t (sec)
Design 1	66	80	90	100	90	100	110	140
Design 2	68	83	70	96	72	94	101	126
Design 3	70	85	70	95	72	95	95	121

parameter for urban planners who seek insights for designing cities with AV influx. The research also focused on strategic planning for future AV influx and making metropolitan cities more amiable to AVs, especially with a focus on the effect of AV penetration in crowded areas like public transport bus bays right outside local train stations. Such locations see the highest influx of daily commuters, and the study proposed designs for aiding public transport with AV as an added option. It can also be concluded that the AV has the promising potential of assisting the public transport system by reducing the travel time of passengers using this commute option. The research results can be enhanced in the future by simulating for infrastructure changes like exclusive road lane for AVs, priority AV signals, and modifying the number of overall lanes in the link.

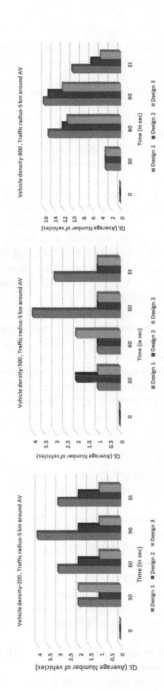

FIGURE 5.12
Queue length with different vehicle density and traffic radius of 5 km around AV.

References

1. "The road to seamless urban mobility", McKinsey & Company, 2019. Available: www.mckinsey.com/Business-Functions/Sustainability/Our-Insights/The-ro ad-to-seamless-urban-mobility

2. R. Bishop, *Intelligent Vehicle Technology and Trends*, Artech House, Norwood, USA 2005.

3. R. Naumann, R. Rasche, J. Tacken, "Managing autonomous vehicles at intersections", *IEEE Intelligent Systems*, vol. 13, no. 3, pp. 82–86, May 1998.

4. Werner, J., (2003), Inside the USDOT's "Intelligent Intersection" Test Facility, ITS Cooperative Deployment Network, Retrieved from: http://ntlsearch.bts.gov/tris/record/tris/00961269.html.

5. A. Stocker and S. Shaheen, "Shared automated vehicles: Review of business models PDF Logo", 2017. Available: http://hdl.handle.net/10419/194044.

6. D. Fagnant, K. Kockelman, "Preparing a nation for autonomous vehicles: Opportunities, barriers and policy recommendations", *Transportation Research Part A: Policy and Practice*, vol. 77, pp. 167–181, 2015. doi: 10.1016/j.tra.2015.04.003.

7. J. Meyer, H. Becker, P. Bösch, K. Axhausen, "Autonomous vehicles: The next jump in accessibilities?", *Research in Transportation Economics*, vol. 62, pp. 80–91, 2017. doi: 10.1016/j.retrec.2017.03.005.

8. L. Hobert, A. Festag, I. Llatser, L. Altomare, F. Visintainer, A. Kovacs, "Enhancements of V2X communication in support of cooperative autonomous driving", *IEEE Communications Magazine*, vol. 53, no. 12, pp. 64–70, 2015. doi: 10.1109/mcom.2015.7355568.

9. D. Dolgov, S. Thrun, M. Montemerlo, J. Diebel, "Path planning for autonomous vhicles in unknown semi-structured environments", *The International Journal of Robotics Research*, vol. 29, no. 5, pp. 485–501, 2010. doi: 10.1177/0278364909359210.

10. T. Teja, "Autonomous robot motion path planning using shortest path planning algorithms", *IOSR Journal of Engineering*, vol. 3, no. 01, pp. 65–69, 2013. doi: 10.9790/3021-03116569.

11. J. Sun, G.Baciu, X. Yu, M. Green, "Template-based generation of road networks for virtual city modeling", in *Proceeding VRST '02 Proceedings of the ACM Symposium on Virtual Reality Software and Technology*, pp, 33–40, Hong Kong, November 2002.

12. H. Wang, J.K. Kearney, J. Cremer, P. Willemsen, "Steering behaviors for autonomous vehicles in virtual environments", in *IEEE Proceedings. VR 2005, pp. 155–162 Virtual Reality*, March 12 2005, Atlanta, USA.

13. S. Donikian, "VUEMS: A virtual urban environment modeling system", in *CGI '97 Proceedings of the 1997 Conference on Computer Graphics International*, June 23–27, 1997, Hasselt and Diepenbeek, Belgium.

14. L. Lefebvre, P. Poulin, "Analysis and synthesis of structural textures", in *Graphics Interface 2000. Proceedings of Graphics Interface 2000*, Montréal, QC, Canada, May 2000, pp. 77–86.

15. M. Chen, "Procedural city modeling and night city lighting". http://graphics.lcs.mit.edu/~maxchen/research.html

16. D. Metz, "Developing policy for urban autonomous vehicles: Impact on congestion", *Urban Science*, vol. 2, pp. 33, 2018.

17. J. Parkin, B. Clark, W. Clayton M. Ricci G. Parkhurst, "Autonomous vehicle interactions in the urban street environment: A research agenda", *Proceedings of the Institution of Civil Engineers - Municipal Engineer*, vol. 171, pp. 15–25, 2018.
18. N. Gavanas, "Autonomous road vehicles: Challenges for urban planning in European cities", *Urban Science*, Vol. 3, No. 2, pp. 1–13, 2019.
19. "VTD - VIRES Virtual Test Drive", VIRES, 2020. Available: https://vires.com/vt d-vires-virtual-test-drive/.
20. ROS.org, "Powering the world's robots", 2020. Available: https://www.ros.org/.
21. The Autoware Foundation, Autoware.ai, 2020. Available: https://www.auto-ware.ai/.
22. TCRP Report 19, "Guidelines for the location and design of bus stops", 1996. Available: www.gulliver.trb.org/publications/tcrp/tcrp_rpt_19-a.pdf.
23. Q. Meng, X. Qu, "Bus dwell time estimation at bus bays: A probabilistic approach", *Transportation Research Part C: Emerging Technologies*, vol. 36, pp. 61–71, 2013. doi: 10.1016/j.trc.2013.08.007.
24. R. Z. Koshy, V. Thamizh Arasan, "Influence of bus-stops on urban traffic flow characteristics", *Journal of Transportation Engineering*, vol. 131, no. 8, 2005, pp. 640–643.
25. "Section 4 commuter facilities design requirements", LTA, Singapore. 2019. Available: www.lta.gov.sg/content/ltagov/en.html,.
26. Y. Shen, H. Zhang, J. Zhao, "Integrating shared autonomous vehicle in public transportation system: A supply-side simulation of the first-mile service in Singapore", *Transportation Research Part A: Policy and Practice*, vol. 113, pp. 125–136, 2018. doi: 10.1016/j.tra.2018.04.004.
27. "Standard design road elements", SDRE14-11 BUS 1-4-REV17, LTA, Singapore. https://www.lta.gov.sg/content/ltagov/en.html, 2020.
28. B. Ghosh, J. Dauwels, U. Fastenrath, "Analysis and prediction of the queue length for non-recurring road incidents", in *2017 IEEE Symposium Series on Computational Intelligence*, 2019, pp. 1–8.
29. TCRP Report 19, "Guidelines for the location and design of bus stops", Transportation Research Board National Research Council, 1996.
30. Z. Li, M. Chitturi, D. Zheng, A. Bill D. Noyce, "Modeling reservation-based autonomous intersection control in VISSIM", *Transportation Research Record: Journal of the Transportation Research Board*, vol. 2381, no. 1, pp. 81–90, 2013. doi: 10.3141/2381-10.
31. F., Zhang, R., Lie, G., Wang, H., Wen, H., Xu, J, "Trajectory planning and tracking control for autonomous lane change maneuver based on the cooperative vehicle infrastructure system", *Expert Systems with Applications*, vol. 42, no. 14, pp. 5932–5946, 2015.

6

A Comprehensive Simulation Environment for Testing the Applications of a V2X Infrastructure

Apratim Choudhury, Tomasz Maszczyk, Chetan B. Math,
Hong Li, and Justin Dauwels

CONTENTS

ABSTRACT It is a well-established fact that road transportation is one of the major contributors of greenhouse gases. The increase in the number of light and heavy vehicles in cities every day is providing additional impetus to this predicament. A solution for reducing greenhouse gas emissions would be to "engineer" traffic to decrease congestion. One particular engineering-based solution can be to provide a green light optimized speed advisory (GLOSA). This method involves communicating the remaining phase time to vehicles approaching a junction, which can be used to compute an approach speed that will not require the vehicle to stop. However, the field implementation of this solution would require the installation of communication devices on both traffic signal heads and on vehicles, which would involve considerable cost and policy changes. Therefore, before implementation, it is desirable that any predicted benefits be gauged using alternative means so that the profitability of the installed infrastructure can be justified. We therefore present a novel simulation platform, which combines VISSIM, MATLAB, and NS3 in order to model the various components of a V2X-based application, such as GLOSA, with a high level of fidelity. In this chapter, we apply this simulation platform to compute the benefits of implementing GLOSA on two intersections, one located in Singapore and the other located in Eindhoven in the Netherlands. The results illustrate the reduction in fuel consumption and queue length due to the execution of GLOSA for different traffic flows, communication parameters, levels of V2X penetration, and traffic signal policy.

6.1 Introduction

It is predicted that the issues pertaining to traffic management and traffic efficiency can be greatly mitigated with the presence of a vehicle-to-everything (V2X) communication infrastructure. A V2X communication network is a vehicular ad hoc network (VANET) that enables vehicles to communicate with each other and with road-side infrastructure units that are within a certain maximum distance from each other. This network allows access to real-time traffic information that can be leveraged to allow the implementation of sophisticated traffic control and vehicle routing algorithms. However, setting up an infrastructure of such a magnitude would involve a sizeable fiscal investment and substantial policy changes. Therefore, before a traffic management strategy built upon a V2X foundation can be implemented on a wide scale, a phase of thorough testing and verification is highly warranted to ensure that the implementation can indeed bring about benefits in traffic movement. Conducting field tests will be expensive and unrealistic in the early stages so the only feasible way of obtaining the relevant data is by carrying out exhaustive simulations. The simulations will need to recreate traffic scenarios as accurately as possible in addition to modeling the communication network with a high level of fidelity so that results derived can be deemed realistic. In order to address this need, the authors have developed a simulation platform that combines three software packages: VISSIM (traffic modeling), MATLAB (traffic management application), and NS3 (communication network simulation).

In metropolitan cities, traffic congestion is emerging as one of the major challenges that need to be overcome [1]. Every day, a growing number of vehicles are being added to the existing fleet, putting more stress on the already overburdened transportation infrastructure. Traffic congestion not only results in wasted hours for commuters, but it also affects their health and well-being [2]. In addition, according to Barth et al. [3], driving patterns typical of congestion, involving frequent acceleration and deceleration in succession, lead to higher CO_2 emissions than if traffic was moving at steady-state speeds, even if the average speed is equal for both cases. In order to address these issues, policy-makers are concentrating on approaching the problem at a microscopic level. Their approach is to push car manufacturers to incorporate improvements in their vehicles. Such improvements include efforts to make vehicles lighter, enhancing powertrain efficiency, and the use of alternative fuels. However, these technologies still have a long way to go before full implementation and market penetration. In order to bring about an effective reduction in emissions, it is important that the root cause be addressed – the intensification of congestion in major cities.

There are already some intelligent traffic management systems in existence around the world. In Singapore, the green link determining system (GLIDE) [4, 5] is in operation. This system is an adaptation of the Sydney coordinated adaptive traffic system (SCATS) [6] which applies real-time traffic information to establish a signal scheduling policy. In addition, there are similar systems currently in place in Singapore, such as TrafficScan [7], junction electronic eyes (J-EYES), the expressway monitoring advisory system (EMAS) [5], and electronic road pricing (ERP) [19]. However, most global traffic management solutions are reactionary systems that are activated only when congestion is observed on any part of the traffic network. A more proactive system to tackle congestion would be to implement a strategy that brings about uniformity in the traffic movement profile, such as the green light optimized speed advisory (GLOSA) system [8]. Therefore, the aim of this work is to utilize the integrated simulation environment [9] consisting of VISSIM (traffic simulation), MATLAB (V2X application modeling), and NS3 (communication simulation) to analyze as exhaustively as possible whether the application of GLOSA leads to any benefits with regard to fuel consumption and queue length. GLOSA simulations are carried out with both a fixed-time traffic signal policy and also the GLIDE system in order to compare the differences in the effects that GLOSA may have. Our aim with this chapter is to demonstrate the capability of the simulation platform to model and assess various facets of a large-scale V2X application. To illustrate the versatility, we obtain results by adding variation to the following traffic and communication parameters:

1) Traffic parameters
   a) Traffic network
   b) Vehicle volume

    c)   Traffic signal timing policy

    d)   V2X penetration rate

2) Communication parameters

    a)   Transmitter power

    b)   Receiver energy detection threshold

    c)   Data rate

Also, as far as we know, GLOSA for GLIDE has not yet been implemented in a testbed or simulation platform. Moreover, the benefits of GLOSA have not yet been assessed in a realistic traffic network.

A precursor to the widespread implementation of GLOSA is the existence of a V2X (with vehicle-to-vehicle/vehicle-to-infrastructure as subsets) communication foundation [10]. Over the past decade, the area of vehicular communication has received much attention, mainly with the aim of improving safety [11] and efficiency on roads [12]. V2X communication typically utilizes the IEEE 802.11p protocol because it allows communication outside the context of the basic service set. This leads to lower latency in communication, making V2X highly suitable for safety-critical applications. GLOSA will leverage on V2X communication technology to advice drivers of optimal driving speeds. However, the installation of V2X communication units across entire road networks of a city and on-board vehicles will involve high costs and policy changes. Hence, any predicted benefits of installing them will need to have convincing quantitative justification. In this regard, the tool can potentially be utilized to test the optimal placement configuration of road-side units (RSUs) and levels of penetration for vehicle on-board units (OBUs).

The chapter is organized as follows. In Section 6.2.1, a review of existing V2X simulation platforms and their shortcomings has been presented. In Sections 6.2.2 and 6.2.3, we discuss congestion mitigation using traditional intelligent transport systems (ITS) and V2X-based techniques, while in Section 6.2.4 we review the research work carried out toward conception and testing of GLOSA. In Section 6.3, we describe the design of the simulation environment, with the parametric setup detailed in Section 6.4. We present our results in Section 6.5. We offer concluding remarks and ideas for future research in Section 6.6.

6.2 Related Work

6.2.1 V2X Simulation Platforms

In Table 6.1, we provide an overview of some of the major simulation environments for testing V2X protocols and applications, along with the shortcomings that motivated us to develop our own simulation environment.

TABLE 6.1

V2X Simulation Platforms

Simulation platform	Advantage	Disadvantage
VSimRTI [13]	Ability to integrate multiple traffic and communication simulators. E.g., SUMO (traffic), JiST/SWANS (communication), MATLAB CCMSim (communication), and OMNET++ (communication)	No interface for VISSIM
Multiple simulator Interlinking for IVC [14]	Integrates VISSIM, MATLAB, and NS2	Not upgraded for NS3
iTetris [15]	Integrates SUMO and NS3 using an iTetris control system middle-ware that promotes extendibility of the architecture	No provision to integrate VISSIM
Veins [16]	Integrates SUMO with OMNET++	No provision to integrate VISSIM
NCTUns 6.0 [17]	Highly integrated traffic and network simulator	Limited flexibility to integrate third-party simulators
TraffSim [18]	Highly detailed traffic simulation and fuel consumption model	No extensions for communication simulation

An important requirement that any given simulation tool needed to fulfill was the usage of VISSIM for traffic simulations. Our decision to implement the intersection models in VISSIM can be attributed to the following reasons: (1) it is able to accurately replicate the trajectories of different classes of vehicles in simulation; (2) allows a simulation resolution of 50 ms, which would enable testing V2X applications that require split-second responses from the vehicle; (3) it has built-in models for fuel consumption and queue length calculation; and (4) it provides real-time data exchange with external programs and allows modification of vehicle behavior based on stimuli provided through the same programs.

For a more detailed review, we refer the reader to [9].

6.2.2 Existing Traffic Congestion Mitigation Methods

Different cities have adopted a number of techniques for traffic congestion management. As two illustrations, we will briefly discuss Singapore and the Netherlands. In Singapore, ITS-based solutions such as GLIDE [4, 5], TrafficScan [7], J-EYES, EMAS [5], and ERP [19] are already in place.

Similarly, in the Netherlands ITS solutions are heavily applied for congestion management and control. Traffic is constantly monitored by the Dutch traffic control center by means of cameras and speed loops. The Dutch traffic control center also uses urban traffic optimization by integrated automation (UTOPIA)/system for priority and optimization of traffic (SPOT) [20], an adaptive traffic control system responsible for automatically determining and implementing optimum management strategies. Table 6.2 summarizes the existing traffic congestion mitigation strategies in both Singapore and Eindhoven.

One issue with the above-mentioned congestion avoidance techniques is that they come into effect only after a congestion situation has transpired. Any mitigatory solution cannot be implemented on a moment's notice and, moreover, it takes a certain amount of time before the congestion dissipates and traffic becomes smooth again. Therefore, research has been focused on coming up with solutions that are not only more proactive but are based on data collected from practical sources (loop detectors, cameras, V2X, etc.). In the next subsection, we will elaborate on the congestion mitigation strategies that rely on a V2X infrastructure.

6.2.3 V2X-Based Traffic Congestion Mitigation

V2X communication has the potential to improve traffic efficiency considerably [21]. Implementation of a pervasive V2X infrastructure will allow vehicle and traffic management centers to access copious amounts of real-time traffic data that will encourage the conception of more sophisticated congestion mitigation solutions. In academia, a lot of effort is being put into developing V2X-reliant optimal route guidance and navigation solutions to combat congestion. Table 6.3 provides a summary of the work currently being done along with the advantages, disadvantages, and their choice of the simulation tool. In the remainder of this section, we will provide more details about the different studies.

TABLE 6.2

Existing Congestion Mitigation Strategies in Singapore and Eindhoven

	Strategy	Principle	Disadvantage
Singapore	ERP	Congestion pricing	Reactive system coming into effect only when a congestion incident occurs
	GLIDE	Traffic smoothing using signal control	
	TrafficScan, J-EYES, and EMAS	Incident management	
Eindhoven	*Groene Golf*	Traffic smoothing using signal timing coordination	
	UTOPIA/SPOT	Traffic smoothing using adaptive signal control	

TABLE 6.3

V2X-Based Congestion Mitigation

Research work	Authors	Disadvantage	Simulation tool
V2X-based traffic congestion recognition and avoidance	Wedel, Jan W., Bjo¨rn Schu¨nemann, and Ilja Radusch [12]	May not work in case of high-density traffic due to data drops	VSimRTI
Reducing traffic jams via VANETs	Knorr, Florian, Daniel Baselt, Michael Schreckenberg, and Martin Mauve [23]	May not work in case of high-density traffic	JiST/SWANS
Real-time traffic congestion management and deadlock avoidance for VANETs	Hussain, Syed Rafiul, Ala Odeh, Amrut Shivakumar, Shalini Chauhan, and Khaled Harfoush [24]	No estimate of algorithm performance for different levels of V2X penetration	Self-built using Visual Studio 2010
Increased traffic flowthrough node-based bottleneck prediction and V2X communication	Backfrieder, Christian, Gerald Ostermayer, and Christoph F. Mecklenbra¨uker [25]	No estimate of algorithm performance for different levels of V2X penetration	TraffSim

Wedel et al. [12] conceptualize the application of V2X for real-time information sharing about the traffic conditions. The technique involves applying Dijkstra's algorithm [22] by viewing intersections and road segments as nodes and edges. However, this strategy comes into effect only when a congestion situation has already occurred. In order to simulate the effectiveness of this algorithm, the authors used V2X simulation runtime infrastructure (VSimRTI), a simulation architecture that is inspired by Institute of Electrical and Electronics Engineers (IEEE) standard for modeling and simulation high-level architecture. The results show that the application of this approach leads to a decrement in travel time by almost 50%, as the V2X penetration rate reaches 80% and higher.

In another congestion avoidance approach introduced by Knorr et al. [23], advisories are provided to the drivers to maintain a larger gap from the preceding vehicle, in case congestion is detected. This helps in the reduction of perturbations that add to the congestion. For the purpose of testing, both traffic and communication have been modeled on the java in simulation time (JiST)/scalable wireless ad hoc network simulator (SWANS) simulators [26]. However, the approach proposed in [23] may not be effective in highly-congested traffic situations because the message dissemination is carried out by means of periodic broadcasts. This may lead to a broadcast storm and collisions when multiple vehicles are broadcasting all at once [27].

Syed Rafiul Hussain et al. [24] have introduced a new protocol named congestion management and deadlock avoidance (COMAD). Approaching vehicles broadcast periodic messages indicating their arrival to a particular traffic intersection, which are then processed to evaluate the average queue length and waiting time. In case of potential congestion, all the neighboring intersections are informed so that they advise alternate routes to the approaching vehicles, to avoid exacerbation of the congestion.

Another approach for congestion avoidance aims to leverage data from vehicle-to-infrastructure (V2I) communication to predict traffic congestion [25]. The novelty of this work lies in the usage of traffic prediction algorithms to identify the possibility of future congestion, in the case of which alternative routes for approaching vehicles can be calculated. Subsequently, infrastructure-to-vehicle (I2V) communication is used for communicating alternative routes to approaching vehicles.

However, in both [24] and [25], it has not been investigated how different penetration levels of V2X technology would affect the protocols and algorithms as all vehicles are assumed to have V2X technology. This is because the simulation environment considered in each approach lacks the capability of carrying out realistic network simulations. Hussain et al. [24] assume ideal physical (PHY) and medium access control (MAC) layers in the simulations; therefore, the effect of path loss, interference, and packet collisions have not been taken into account.

In the next section, we go into greater detail behind the conception of GLOSA and the studies that have been carried out to prove the effectiveness of this idea.

6.2.4 Green Light Optimized Speed Advisory

In this chapter, we aim to simulate the V2X-enabled GLOSA and investigate the benefits that it can bring to traffic flow efficiency. However, before analyzing the results, we present a brief review of the research carried out in developing GLOSA as well as the simulation tools used for testing. Table 6.4 summarizes representative research studies on GLOSA, the simulation tools applied in those studies, and the shortcomings of each tool.

The concept of GLOSA was first conceived in 1983 when Volkswagen [30] introduced the Wolfsburger Welle. The idea was to communicate speed advisory using infrared technology. However, the project was discontinued due to technical issues and a lack of commercial acceptance. Interest in GLOSA resurfaced when the IEEE 802.11p communication protocol got introduced, as this opened up a new avenue for data exchange. Some of the projects that capitalized on dedicated short range communication (DSRC) for implementation of GLOSA were the Travolution project [31], simTD [32], and PREDRIVE C2X [33].

Raubitschek et al. [29] investigated how fuel consumption can be reduced if the traffic velocity profile dynamics can be pre-adjusted based on an upcoming traffic event, such as a traffic signal. Simulations and real-life experiments showed a reduction of 13% in fuel consumption (and CO_2 emissions).

TABLE 6.4

Simulation Tools Utilized in GLOSA Research

Research work	Simulation tool	Simulation shortcomings
Impact of signal phasing information accuracy on GLOSAs [8]	Traffic simulation – **VISSIM** Communication simulation – **VISSIM Car2X**	Car2X has been discontinued by VISSIM
Performance study of a GLOSA application using an integrated cooperative ITS simulation platform [28]	Traffic simulation –**SUMO** Communication simulation – **JiST/SWANS**	Only fixed-time traffic signal can be modeled
Predictive driving strategies under urban conditions for reducing fuel consumption based on vehicle environment information [29]	Vehicle and predictive trajectory simulations in **MATLAB**	No traffic simulation and perfect communication assumed

However, only a single traffic light was investigated and the velocity adjustments are made assuming vehicles do not influence each other, which is highly unlikely.

Katsaros et al. [28] tested a GLOSA-like scenario in a two-intersection model developed in SUMO (a road traffic simulation suite). The authors incorporate JiST/SWANS to carry out network simulations and applied the Fraunhofer VSimRTI to enable an online interface between traffic and communication simulation models. The results show that higher GLOSA penetration rates provide more benefits in terms of fuel consumption and traffic congestion. Also, the optimal activation distance where the GLOSA application should advise the driver is found to be approximately 300 m from the traffic lights. Even though a more accurate communication model was utilized, the analysis was done for only a fixed-time traffic signal.

In order to make the simulations more realistic, Stevanovic et al. [8] take into consideration the queue discharge time in their calculation of the optimized speed advisory. However, the communication simulations were carried out using the Car2X module of VISSIM, which is no longer an available feature.

Weisheit et al. [34] and Bodenheimer et al. [35] are two of the few research studies that have attempted to implement GLOSA for an adaptive traffic light scheduling policy. However, in both cases, before an advisory is broadcast, the GLOSA system will need to predict the dynamically assigned phase durations, which may or may not be accurate. In the case of GLIDE, cycle times and green times are calculated for each cycle based on the volume of traffic observed in the previous cycle and this new time does not change for the entire cycle duration. Therefore, it is possible to know the phase change time in advance. Since we are modeling the GLIDE system, our study will not be affected by the aforementioned predicament.

In summary, as can be observed from Sections 6.2.3 and 6.2.4, a number of traffic congestion strategies have been conceived that depend heavily on data acquired or disseminated using V2X communication. In order to test their approaches and algorithms, the researchers have resorted to either a self-developed simulation tool or one of the tools referred to in Section 6.2.1. However, as the research typically concentrated on either the traffic or the communication aspect, detailed modeling has been carried out only for the specific area of interest. In the few studies where both the traffic and communication aspects were modeled, the simulation environment lacked the capability to model extra specifics, such as adaptive traffic signals or various levels of V2X penetration. Therefore, we decided to develop a simulation framework that allowed us to address these various factors in the testing of GLOSA. GLOSA has been modeled based on the algorithm developed in [8], and we analyzed how various variables, including the V2X penetration rate, traffic flows, and presence of both fixed time and adaptive traffic signals (GLIDE) affect its performance.

6.3 Simulator Design

In this section, we describe the simulation design and architecture. Figure 6.1 shows the structure of the simulator platform. VISSIM and MATLAB communicate with each other via VISSIM's COM interface, which enables access to most attributes of the traffic simulation, such as vehicle speeds, positions, and signal phase information, through MATLAB. Both VISSIM and MATLAB were installed on a Windows OS since VISSIM only runs on a Windows platform. On the other hand, NS3 has been developed for Linux. To couple VISSIM/MATLAB with NS3, we set up a Linux virtual machine, installed NS3 on it, and linked the host machine (with Windows) to the virtual machine via a virtual network. We then applied socket application programming interfaces (APIs) for the communication of data between MATLAB and NS3.

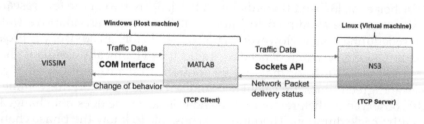

FIGURE 6.1
Block diagram of the simulation environment.

However, in order to design an online simulator for V2X operations, one needs to take note of several other considerations which have been listed as follows:

- Time synchronization between a continuous-time simulation environment (VISSIM) and a discrete-event simulator (NS3)
- Dynamic addition and removal of nodes in NS3
- Choice of appropriate mobility model in NS3

The following subsections elaborate on the strategies that have been applied to address the aforementioned considerations.

6.3.1 Time Synchronization between a Continuous-Time Simulation Environment and a Discrete-Event Simulator

In order to synchronize the three simulators, TCP/IP blocking sockets were utilized. Once NS3 finishes its initial configuration, it enters the simulation phase, upon which the code is blocked from proceeding at the zeroth second. Subsequently, the traffic simulations are triggered via a COM API function. As soon as the first group of vehicles enters the network, their positions are relayed to NS3 along with the time at which they enter the network. This allows scheduling a position update event in the network simulator at the same virtual time as VISSIM. Simultaneously at this point, VISSIM enters a blocked state and needs to wait till NS3 reverts back with the results of a communication simulation. Blocking events are scheduled every simulation second for obtaining data from MATLAB, which ensures that at every subsequent second of VISSIM virtual time, appropriate vehicle data is communicated to NS3. A flowchart describing this process has been presented in Appendix A.

6.3.2 Dynamic Addition and Removal of Nodes in NS3

In NS3, new nodes cannot be created nor existing nodes be destroyed until the end of the simulations. In order to work around this, a fixed number of nodes are created during the NS3 configuration phase without any virtual sockets attached. Sockets are only attached to the appropriate number of nodes during runtime, as there are vehicles in the network. In order to characterize a vehicle in NS3, a *vehicle node* structure is created with the following fields:

- `Ptr<Node> veh _ node` : Smart pointer to the NS3 node object
- `Ptr<Socket> veh _ socket` : Smart pointer to the NS3 socket attached to the node
- `int veh _ number` : Integer vehicle number
- `bool in _ use` : Boolean true or false to indicate if the node is currently in use

A vector of such *vehicle node* structure objects is created, and each NS3 node is then loaded onto the veh _ node member of the object. Every time a new vehicle enters the network, it will be associated with a *vehicle node* object, which is currently not "in use", and a socket will be attached. Once a *vehicle node* is active, only its mobility model is updated every simulation second. When the vehicle leaves the network, communication is disabled by closing the socket and setting the in _ use parameter to *false*. In order to use the same node object for a different vehicle, we bind the socket to an address and re-enable *Recv* calls on it. Algorithm 1 (below) presents the pseudocode to achieve the aforementioned process.

6.3.3 Choice of Appropriate Mobility Model in NS3

In order to model the mobility of the vehicles accurately, we chose the way-point mobility model in NS3. To provide waypoints in an online manner, we decided to run VISSIM, one simulation step in advance, as has been described in [36]. This ensured that both origin and destination waypoints are always available to simulate node mobility.

6.4 Simulation Setup

6.4.1 Traffic Network

For the purpose of our study of GLOSA, we have modeled an intersection in Singapore, as shown in Figure 6.2 and one in Eindhoven, Netherlands,

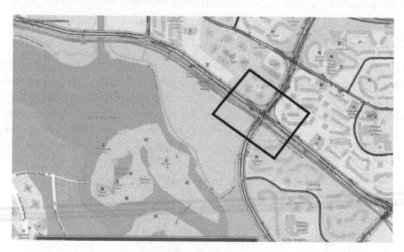

FIGURE 6.2
VISSIM representation of the traffic network in Singapore.

Figure 6.3. There is no specific reason to choose these two particular intersections, apart from the purpose of demonstrating the applicability of the simulation tool to any traffic network, for both left-hand drive (Singapore) and right-hand drive (Eindhoven) traffic. The traffic simulation has been carried out over the entire network depicted in Figures 6.2 and 6.3. However, data regarding fuel consumption and queue length have been collected from within the nodes marked with black borders. We have not modeled traffic signals on the three minor intersections in the north-west, north-east and the south-east approaches of the Eindhoven network. We assume that there is a thoroughfare all the way from the start of the links, as we wanted the approaches to the intersection to be at least 200 m in length. All the traffic flow values referred to in Figures 6.5, 6.6 and 6.7 correspond to the total vehicle flow on all the links inside the nodes.

We also assume that all traffic signal heads in the two intersections have an RSU attached to them. These RSUs broadcast the following information at 1 Hz: the signal head location, current phase, and the remaining phase time of the traffic signals.

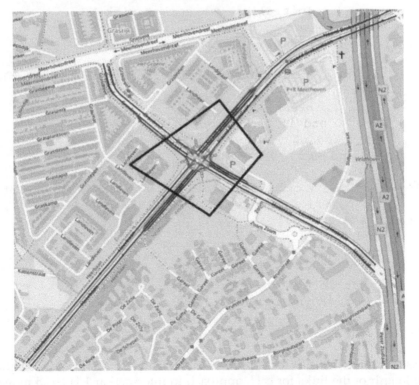

FIGURE 6.3
VISSIM representation of the traffic network in Eindhoven, Netherlands: (a) Singapore phase plan, (b) Eindhoven phase plan.

Algorithm 1: Dynamic addition and removal of nodes

```
Requires:
  A vector [N_i], i = 1, 2, . . . , n, of pre-configured
  NS-3 nodes
  A vector [V node_m], m = 1, 2, . . . , n, of Vehicle_
  Node structures

1: Initialize:
     [V node_m].veh_node ←[N_i]
2: for t = 1 to T do
3:      Receive list of vehicles [V L_t] and their
        positions from VISSIM
4:      for v ∈ [V L_t] do
5:         if v / [V node_m] then
6:            for k = 1 to n do
7:               if V node_k.in_use = False then
8:                  V node_k ← v; t> Record veh_number and set
                                       in_use to True.
9:                  if veh_socket ⇒NULL then
10:                    Initialize veh_socket with NS-3
                       socket object;
11:                    break;
12:                 else
13:                    Bind address to veh_socket and re-enable
                       Recv flag;
14:                    break;
15:                 end if
16:              end if
17:           end for
18:        else
19:           Update mobility model of v using its current
              position received from VISSIM;
20:        end if
21:     end for
22:     for v_n ∈([V L_t] ∩[V node_m])^c do
23:        Close veh_socket of v_n;
24:        Set in_use of v_n to False;
25:        Set veh_number of v_n to 0;
26:     end for
27: end for
```

Each intersection has four approaches. For the intersection in Singapore, the length of the links for each approach to intersection 1 is 603.5 m (east-to-west link), 706 m (west-to-east link), 250 m (north-to-south link), and 296 m (south-to-north link), respectively. Similarly, the length of links on the Eindhoven intersection is 720 m (south-west to north-east), 560 m (north-east to south-west), 350 m (north-west to south-east) and 553 m (south-east to north-west), respectively. The length of the links is important since the activation distance (distance at which a vehicle first receives information from the RSU) has a strong impact on the effectiveness of GLOSA [28].

The information broadcast from a particular signal head is only relevant when the vehicle is on the link served by the traffic signal. Therefore, even if the vehicle is within range, and consequently receives a packet, those packets will be discarded if the vehicle is not on the corresponding link. In order to encode the association between a traffic signal and its corresponding link, we create a mapping between the link number and the MAC address of the RSU attached to the signal head.

6.4.2 Signal Timing

Signal timing for the Singapore intersection: Since the Singapore traffic signal system is governed by GLIDE, we decided to model the GLIDE algorithm following the approach of Daizong et al. [37]. They developed a SCATS-based API plug-in for the Paramics traffic simulator derived from GLIDE output files provided by the Land Transport Authority of Singapore (LTA). The base cycle time is chosen to be 100 s, and the signal phase plan has been illustrated in Figure 6.4(a). New cycle times are generated by adding/subtracting a few seconds to the previous cycle time, with the maximum being 6 seconds. However, if for three consecutive cycles, the cycle time is increased/decreased by 6 seconds, then the maximum switches to 9 seconds. In Singapore, the maximum value the cycle time can be increased to is 140 seconds, while the minimum is 60 seconds. For more details on the GLIDE emulation algorithm and the signal phase split choices, we refer the reader to [37].

Signal Timing for the Eindhoven intersection: We assume that the Eindhoven traffic intersection is governed by a fixed-time signal policy. A Protected-Permissive Left-Turn Phasing has been considered, as shown in Figure 6.4(b). The fixed cycle time has been set

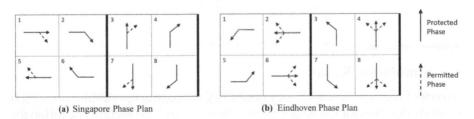

(a) Singapore Phase Plan   (b) Eindhoven Phase Plan

FIGURE 6.4
Ring barrier diagrams for Singapore and Eindhoven.

to 100 seconds while the individual phase times are 15 seconds, 35 seconds, 15 seconds, and 25 seconds for phase 1&5, phase 2&6, phase 3&7, and phase 4&8, respectively. For both the signal groups, the amber time is 3 seconds, with a 2 second all-red between change of phase between the groups.

We compute the GLOSA advisories from the timings of signal phases 2&6 and 4&8, since the left turn is permissive, and the straight-going vehicles will be the main beneficiaries of GLOSA.

6.4.3 Fuel Consumption Modeling

We applied the VISSIM node evaluation module for determining fuel consumption. The module utilizes the same formulas that are present in the traffic signal system simulation and optimization software TRANSYT 7-F for calculating fuel consumption of vehicles [38, 39]. TRANSYT 7-F allows the optimization of signal times with respect to delay, the number of intersection stops, and fuel consumption. The underlying model of fuel consumption is a function of: (1) total travel (vehicle-mi/hr), (2) total delay (vehicle-hr/hr), (3) total stops (full stops/hr), and (4) free speed on each link (mph).

The model was calibrated on data of fuel consumption gathered by the Oak Ridge National Laboratory of the US Department of Energy [40] in 1985. The tests combined laboratory and on-road experiments with 15 vehicles, and the data was collected as a function of both acceleration and velocity.

6.4.4 Communication Simulation

For the purpose of simulation, we have utilized NS3 because of the justifications as elaborated in [41]. In the network simulations, the signal head RSU node is programmed to broadcast 500-byte-long messages at 1 Hz (1 messages/second). Four different combinations of transmit power, energy detection threshold, and data rate have been chosen to incorporate variation in the activation distance and analyze the effects on fuel consumption. Table 6.5 captures four different settings of the aforementioned communication parameters associated with an increasing activation distance.

6.5 Simulation Results

In order to determine the efficacy of GLOSA, we conducted simulations of 800 seconds duration for the Singapore intersection and 600 seconds duration for the Eindhoven intersection. We added an extra 200 seconds to the simulation time on the Singapore intersection to generate historical data for the GLIDE

TABLE 6.5

Combinations of Communication Parameters

Parameter Table	
Communication parameters	**Parameter value**
Transmit power	12 dBm
Energy detection threshold	−70 dBm
Data rate	9 Mbps
Transmit power	15 dBm
Energy detection threshold	−75 dBm
Data rate	9 Mbps
Transmit power	23 dBm
Energy detection threshold	−87 dBm
Data rate	6 Mbps
Transmit power	28.8 dBm
Energy detection threshold	−95 dBm
Data rate	6 Mbps

algorithm. The minimum and maximum advisory that GLOSA can hand out to a vehicle is assumed to be 20 km/hr and 100 km/hr, respectively. While there is no minimum speed limit, we chose the lowest assigned speed to be 20 km/hr so that a slow speed does not lead to congestion. Furthermore, we are interested in assessing the effectiveness of GLOSA for different traffic volumes. To this end, we varied the vehicle densities to create conditions for mild to high congestion. We also performed simulations for V2X penetration rates of 0%, 10%, 20%, and 100%.

For both the Singaporean and Dutch intersections, it can be observed that for low RSU power values and high data rates (hence low activation distance), the effects of GLOSA are not favorable for either low- or high-density traffic. These results exhibit a trend similar to the results of [28], in which the correlation between GLOSA effectiveness and activation distance has been analyzed. In addition, for the Singaporean network, the negative effects are exacerbated due to lack of information about the cycle time in the subsequent traffic signal cycles, for both low and high power RSUs. This affects vehicles that cannot make it through the green phase of the current cycle and have to cross in the green phase of the next cycle but have calculated the advisory speed based on the phase timings of the current cycle. By increasing the RSU power and reducing both the data rate and the energy sensitivity of the receiver, the activation distance increases, and consequently GLOSA performs better even with high traffic volumes. However, the decrease in fuel consumption and queue length, even for higher activation distances, is not consistent for the Singapore network due to the aforementioned restrictions posed by the GLIDE system.

Each curve in Figures 6.5 and 6.6 shows the fuel consumption for various V2X penetration rates and traffic volumes. For the Singaporean intersection in Figure 6.5, with a transmit power of 23 dBm, the data rate of 6 Mbps,

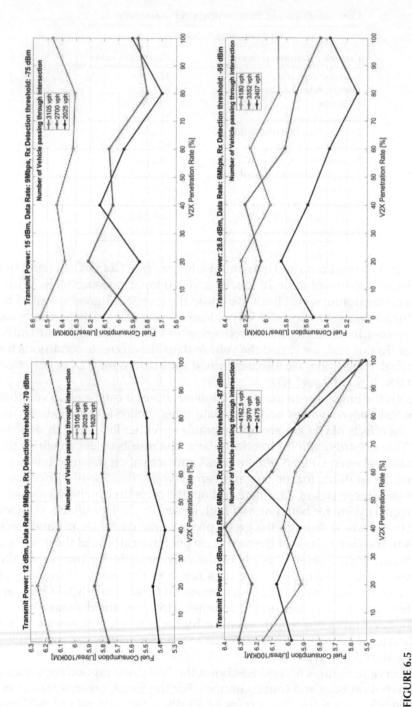

FIGURE 6.5
Fuel consumption for the Singapore intersection.

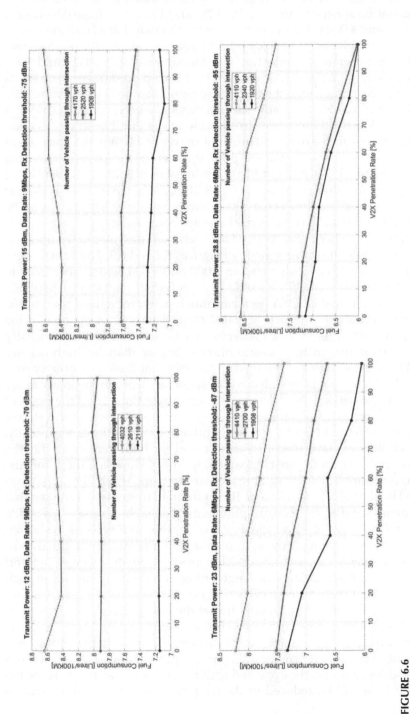

FIGURE 6.6

Fuel consumption for the Eindhoven intersection: (a) queue length statistics on the Singapore network, (b) queue length statistics on the Eindhoven network.

and receiver sensitivity of −87 dBm, the fuel reductions from no V2X to full V2X penetration are approximately 3%, 12%, and 4.7% for vehicle densities of 2,500, 3,000, and 4,000 vehicles per hour approximately. For a transmit power of 28.8 dBm, the data rate of 6 Mbps, and receiver sensitivity of −95 dBm, the fuel consumption minimization is approximately 2.5%, 11%, and 5.3% for vehicle flows of 2,400, 3,300, and 4,100 vehicles per hour. For each set of communication parameters, the fuel consumption value is different for the no–V2X scenario, since the traffic simulations have been conducted with a different random seed each time. In the case of the Eindhoven intersection in Figure 6.6, we can see that for a transmit power of 23 dBm, the data rate of 6 Mbps, and energy detection of −87 dBm, the fuel consumption decrements are approximately 18%, 12%, and 10.5% for densities of 2,000, 2,700, and 4,400 vehicles per hour, while for a power of 28.8 dBm, the data rate of 6 Mbps, and sensitivity of −95 dBm, the decrements are in the order of 16%, 18%, and 9% for 2,000, 2,400 and 4,200 vehicles per hour approximately.

As mentioned earlier in Section 6.4.1, the vehicle flows refer to the flows on all the links inside the polygon node in Figures 6.2 and 6.3. The vehicle densities have been set randomly between 2,000 and 4,000 vehicles per hour to analyze the effects of GLOSA for mild to high traffic congestion. Therefore, it can be observed that GLOSA performs much better for a fixed time traffic signal control than for a vehicle actuated control system. In the case of the fixed time traffic signal, we can observe that for low and medium density traffic, the reduction in fuel consumption is higher than for high-density traffic. This can be mainly attributed to the fact that there is a greater proportion of vehicles that is provided with a speed advisory in the lower range of speeds, i.e., 20–40 km/hr. These speeds are not optimal from the point of view of fuel consumption, and due to a greater distribution of the slower vehicles, they are in a better position to influence the traffic. Consequently, the overall flow of traffic becomes slower and for high-density traffic, the fuel consumption benefits are comparatively lower. However, the same justification is not applicable to the Singaporean network, where the signals are governed by the vehicle actuated GLIDE system. In this case, we can see that for low-density traffic, fuel consumption reduction is the lowest. This is due to the fact that if a reduced vehicle flow is observed, GLIDE tends to shorten the cycle time and changes the split plan in order to distribute the degree of saturation evenly across all links. This mostly leads to a longer red and a shorter green phase. Due to the lengthening of red phase timings in each cycle, the advisory that is calculated using the previous cycle is not sufficient for the vehicle to arrive at the intersection at the next green phase. Therefore, most of the vehicles, even after receiving an advisory, are forced to come to a stop at the intersection and start from zero once the lights turn green. As a result, this leads to greater fuel consumption. However, for medium and high-density traffic, the cycle and split times either stay the same or the red phase time will be reduced in the next cycle, resulting in the vehicles being able to make it through the intersection without having to stop. For high-density traffic, the reduction in fuel consumption is again less than for medium density traffic for the same reason as for fixed-time traffic signal. For more comprehensive results, the traffic model can be calibrated with the relevant traffic flow data, for any road network of interest. In addition, one can also accurately model the V2X communication channel characteristics by integrating a realistic physical layer model into NS3, as has been shown in [42].

In addition to the fuel consumption, we have also shown decrements in queue length due to the implementation of GLOSA in Figures 6.7(a) and 6.7(b). Again, it can be seen that for weaker radio configurations (lower activation distances), the decrements are not very significant. For the Eindhoven intersection in Figure 6.7(b), it can be observed that the implementation of GLOSA is actually increasing queue length. For fixed traffic signals, we see that the queue length reductions for low, medium and high-density traffic are more or less comparable, differently from fuel consumption. Even if a slower speed is assigned to a larger proportion of vehicles (in the high-density case), GLOSA ensures that the vehicles arrive at the intersection at a green phase. Therefore, the reduction in queue length is bound to happen.

Furthermore, from Figures 6.5, 6.6, and 6.7, it can be seen that GLOSA performances for the 23 dBm RSU transmit power and 28.8 dBm transmit power are comparable. This complies with the conclusion that was drawn by [28], that beyond an optimal activation distance, the benefits of GLOSA saturates. Therefore, we can conclude that if GLOSA is to be implemented using V2X technology, it is necessary for the optimum transmission power of the traffic light RSU to be established first. The optimum value will depend on a number of factors, including the length of the road, the topology of the road network, the configuration of the static environment, and the receiver sensitivity. Furthermore, the effectiveness of GLOSA will also depend on the traffic flow at various times of the day, since traffic gridlocks may lead to a reduction in maneuvering space and increased load on the bandwidth. Since it is not feasible to establish the optimal infrastructure placement for all traffic conditions using only field tests, simulations become even more necessary.

6.6 Conclusion and Future Work

In this chapter, we aimed to explore a simulation approach to analyze V2X applications in realistic traffic scenarios, illustrated by the example of V2X-enabled GLOSA. To this end, we chose to simulate the application of GLOSA for two intersections in order to explore various parameters that one can

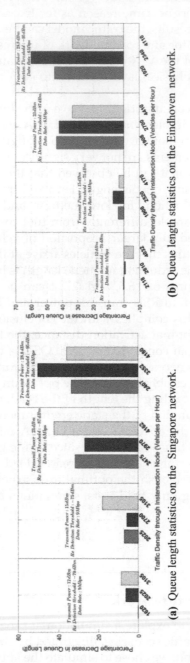

(a) Queue length statistics on the Singapore network.

(b) Queue length statistics on the Eindhoven network.

FIGURE 6.7

Queue length reduction in both networks due to the application of GLOSA.

influence and evaluate using the simulation tool. Specifically, we considered the following parameters:

- Representation of a realistic road network
- Modeling of fixed and vehicle actuated traffic signals (GLIDE) in this case
- Modeling a VANET with different communication parameters (transmit power, receiver sensitivity, data rate, etc.) for RSUs and OBUs
- Ability to incorporate realistic path loss models
- Ability to measure traffic performance metrics, such as fuel consumption
- Analyze the effects of multiple penetration levels of V2X technology
- Study the effects of the different V2X protocols

From the simulations, we observed that the results obtained concur with work done on GLOSA by researchers previously. In addition, the results also indicate that the performance of GLOSA strongly depends on communication parameters. Consequently, the selection of optimal parameters is highly conducive to the success of GLOSA. In most of the previous simulation attempts, the researchers focused on accurately modeling either the traffic component or the communication component of a V2X application since both aspects are different fields of study. To address this shortcoming, we have integrated a realistic traffic simulator with a network simulator, in order to bridge the gap between the two disciplines. In the future, we intend to simulate more applications of V2X, along with integrating more realistic channel models. Such an approach would enable more accurate simulations of V2X applications with a much more accurate model of the V2V physical layer characteristics.

References

1. Cohn, Nick. "TomTom Traffic Index: Toward a global measure." *ITS France.* 2014.
2. Levy, Jonathan I., Jonathan J. Buonocore, and Katherine Von Stackelberg. "Evaluation of the public health impacts of traffic congestion: A health risk assessment." *Environmental Health* 9.1 (2010): 65.
3. Barth, Matthew, and Kanok Boriboonsomsin. "Real-world carbon dioxide impacts of traffic congestion." *Transportation Research Record: Journal of the Transportation Research Board* 2058 (2008): 163–171.
4. Keong, Chin Kian. "The GLIDE system—Singapore's urban traffic control system." *Transport Reviews* 13.4 (1993): 295–305.

5. Chia, Eng Seng. "Engineering Singapore's Land Transport System." In *Complex Systems Design & Management Asia*, Eng Seng Chia (ed.). Springer, Cham, 2015. 99–109.

6. Aydos, J. Carlos, and Andrew O'Brien. "SCATS ramp metering: Strategies, arterial integration and results." In *2014 IEEE 17th International Conference on Intelligent Transportation Systems (ITSC)*. IEEE, Qingdao, China, 2014.

7. Chin, K. K., and C. W. Lee. "TrafficScan-bringing real-time travel information to motorists." Land Transport Authority, Singapore (2009): 7–14.

8. Stevanovic, Aleksandar, Jelka Stevanovic, and Cameron Kergaye. "Impact of signal phasing information accuracy on green light optimized speed advisory systems." In *Transportation Research Board 92nd Annual Meeting*, Washington, DC, 2013.

9. Choudhury, Apratim, et al. "An Integrated Simulation Environment for Testing V2X Protocols and Applications." *Procedia Computer Science* 80 (2016): 2042–2052.

10. Weiß, Christian. "V2X communication in Europe. From research projects towards standardization and field testing of vehicle communication technology." *Computer Networks* 55.14 (2011): 3103–3119.

11. Hafner, M. R., et al. "Automated vehicle-to-vehicle collision avoidance at intersections." In *Proceedings of World Congress on Intelligent Transport Systems*, Orlando, FL, 2011.

12. Wedel, Jan W., Bjo¨rn Schu¨nemann, and Ilja Radusch. "V2X-based traffic congestion recognition and avoidance." In *2009 10th International Symposium on Pervasive Systems, Algorithms, and Networks (ISPAN)*. IEEE, Kaohsiung, Taiwan, 2009.

13. Schu¨nemann, Bjo¨rn. "V2X simulation runtime infrastructure VSimRTI: An assessment tool to design smart traffic management systems." *Computer Networks* 55.14 (2011): 3189–3198.

14. Lochert, Christian, et al. "Multiple simulator interlinking environment for IVC." In *Proceedings of the 2nd ACM International Workshop on Vehicular Ad Hoc Networks*. AC Med, Cologne, Germany, 2005.

15. Rondinone, Michele, et al. "iTETRIS: A modular simulation platform for the large scale evaluation of cooperative ITS applications." *Simulation Modelling Practice and Theory* 34 (2013): 99–125.

16. Noori, Hamed. "Realistic urban traffic simulation as vehicular ad hoc network (vanet) via veins framework." In *12th Conference of Open Innovations Framework Programme*. FRUCT, Oulu, Finland, 2012.

17. Wang, Shie-Yuan, and Chih-Che Lin. "NCTUns 6.0: A simulator for advanced wireless vehicular network research." In *2010 IEEE 71st Vehicular Technology Conference (VTC 2010-Spring)*. IEEE, Taipei, Taiwan, 2010.

18. Backfrieder, Christian, Christoph F. Mecklenbrauker, and Gerald Ostermayer. "TraffSim – A traffic simulator for investigating benefits ensuing from intelligent traffic management." In *2013 European Modelling Symposium (EMS)*. IEEE, 2013.

19. Goh, Mark. "Congestion management and electronic road pricing in Singapore." *Journal of Transport Geography* 10.1 (2002): 29–38.

20. Studer, L., Ketabdari, M., and Marchionni, G. "Analysis of adaptive traffic control systems design of a decision support system for better choices." *Journal of Civil & Environmental Engineering* 5.195 (2015): 2.

21. Baldessari, Roberto, et al. "Ca res-2-car communication consortium manifesto." (2007).
22. Bondy, John Adrian, and Uppaluri Siva Ramachandra Murty. *Graph Theory with Applications*, Vol. 290, Macmillan, London, 1976.
23. Knorr, Florian, et al. "Reducing traffic jams via VANETs." *IEEE Transactions on Vehicular Technology* 61.8 (2012): 3490–3498.
24. Hussain, Syed Rafiul, et al. "Real-time traffic congestion management and deadlock avoidance for vehicular ad hoc networks." In *2013 10th International Conference on High Capacity Optical Networks and Enabling Technologies (HONET-CNS)*. IEEE, Magosa, Cyprus, 2013.
25. Backfrieder, Christian, Gerald Ostermayer, and Christoph F. Mecklenbra"uker. "Increased traffic flow through node-based bottleneck prediction and V2X communication." *IEEE Transactions on Intelligent Transportation Systems* 18.2 (2017): 349–363.
26. Barr, Rimon, Zygmunt J. Haas, and Robbert Van Renesse. "Jist: Embedding simulation time into a virtual machine." In *EuroSim Congress on Modelling and Simulation*, Paris, France, 2004.
27. Milojevic, Milos, and Veselin Rakocevic. "Distributed road traffic congestion quantification using cooperative VANETs." In *2014 13th Annual Mediterranean Ad Hoc Networking Workshop (MED-HOC-NET)*. IEEE, Piran, Slovenia, 2014.
28. Katsaros, Konstantinos, et al. "Performance study of a green light optimized speed advisory (GLOSA) application using an integrated cooperative ITS simulation platform." In *2011 7th International Wireless Communications and Mobile Computing Conference*. IEEE, Istanbul, Turkey, 2011.
29. Raubitschek, Christian, et al. "Predictive driving strategies under urban conditions for reducing fuel consumption based on vehicle environment information." In *2011 IEEE Forum on Integrated and Sustainable Transportation System (FISTS)*. IEEE, Vienna, Austria, 2011.
30. Zimdahl, W. "Wolfsburger Welle–Ein Projekt der Volkswagen Forschung." Volkswagen Forschung, Wolfsburg, 1983.
31. Braun, Robert, et al. "TRAVOLUTION – Netzweite Optimierung der Lichtsignalsteuerung und LSA-Fahrzeug-Kommunikation." *Straßenverkehrstechnik* 53 (2009): 365–374.
32. Stu"bing, Hagen, et al. "Sim TD: A car-to-X system architecture for field operational tests [Topics in Automotive Networking]." *IEEE Communications Magazine* 48.5 (2010): 148–154.
33. PRE-DRIVE C2X project: http://www.pre-drive-c2x.eu
34. Weisheit, T. "Algorithmenentwicklung zur Prognose von Schaltzeitpunkten an verkehrsabha"ngigen Lichtsignalanlagen." *HEUREKA* 14 (2014): 320–339.
35. Bodenheimer, Robert, et al. "Enabling GLOSA for adaptive traffic lights." In *2014 IEEE Vehicular Networking Conference (VNC)*. IEEE, Paderborn, Germany, 2014.
36. Eichler, Stephan, et al. "Simulation of car-to-car messaging: Analyzing the impact on road traffic." In *13th IEEE International Symposium on Modeling, Analysis, and Simulation of Computer and Telecommunication Systems*. IEEE, Atlanta, GA, 2005.
37. Daizong, Liu. *Comparative Evaluation of Dynamic TRANSYT and SCATS-Based Signal Control Systems Using Paramics Simulation*. Dissertation, ScholarBank@ NUS Repository, Singapore, 2003.

38. Penic, Michael A., and Jonathan Upchurch. "TRANSYT-7F: Enhancement for fuel consumption, pollution emissions, and user costs." *Transportation Research Record* 1360 (1992): 104–111.
39. Miller, P., and R. Narayanan. "Microscopic traffic simulation/animation software applications in plnning level transportation decision making." In *Compendium of Papers. District 6 Annual Meeting*. Institute of Transportation Engineers, Nashville, TN, 2000.
40. McGill, R. *Fuel Consumption and Emission Values for Traffic Models*. No. FHWA/RD-85/053. 1985.
41. Choudhury, Apratim, et al. "An integrated V2X simulator with applications in vehicle platooning." In *2016 IEEE 19th International Conference on Intelligent Transportation Systems (ITSC)*. IEEE, 2016.
42. Boeglen, Herve, et al. "A survey of V2V channel modeling for VANET simulations." In *2011 Eighth International Conference on Wireless On-Demand Network Systems and Services (WONS)*. IEEE, 2011.

Appendix A

Flow of Data Amongst Simulators

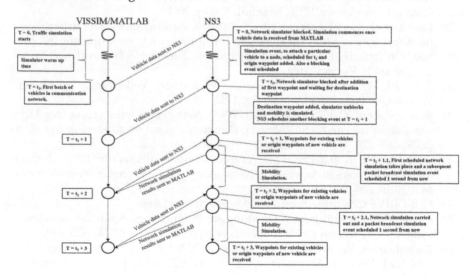

7

A Novel Approach in Communication
Security of Internet-of-Vehicles
Using the Concept of Recurrence
Relation and Symmetric Key

Anirban Bhowmik, Sunil Karforma, Joydeep Dey, and Arindam Sarkar

CONTENTS

ABSTRACT   The current roadsafety statistics in many countries are horrifying. Many people are killed and injured in road accidents. To reduce this problem, governments and manufacturers have launched different initiatives like use of safety belts, airbags, anti-blocking brake systems, and smart vehicular transportation systems. Upcoming traffic safety initiatives in smart transportation systems depend on information technology and this technology also helps to authenticate and trace vehicles in the system.

Recently, smart vehicular system use different types of networks, such as vehicular ad hoc networks (VANETS) and networks based on artificial intelligence, that aim to provide a safer, coordinated, smooth, and smart mode of transportation. This article focuses on communication security issues in smart vehicular applications. The ability to transmit messages efficiently and honestly among vehicles is the key issues in this system. At present, communication in smart transportation system or IoV is vulnerable to various types of security attacks because it uses an open wireless connection. Different types of attacks are described. A vehicle needs to verify the incoming message and authenticate the sender and, in addition, in the dense environment, a vehicle may receive multiple messages at the same time. Therefore, the ability to complete the authentication of multiple messages in a short time is an urgent problem. A huge amount of data is available in the air. So, an unauthorized person may capture data, modify records, or attack vehicular systems. Hackers can access vehicle sensors that control safety devices in a vehicle, shutdown systems, control door locking operations, and disable the car. They could provide false information to drivers, use denial-of-service attacks to bring down the network of vehicles, and they can even download incorrect navigation maps. To address these problems, here we have introduced some techniques using the concept of recurrence relation, hash function, and non-linear function. Our proposed technique is made up of three modules in which first algorithm uses the concept of recurrence relation for random number generation, and this random number is used for session key generation. In the encryption process, the symmetric key, session key, and circular left shift operation are used. The second algorithm uses a hash function for user authentication, and third algorithm uses a decryption process. The different types of experiments on our proposed technique and their results confirm that our scheme is secure, robust, and efficient for data transmission in smart vehicular applications.

7.1 Introduction

Nowadays, road traffic activities are one of the most important issues in daily routines worldwide. Passenger and freight transport are essential for human development and society. In today's digital world, an intelligent transportation system (ITS) plays a vital role in smart cities where citizens' quality of life isimproved. These days, traffic efficiency and controlling are made easy by using ITS. The ITS offers pervasive and robust services in road and traffic safety, reducing traffic congestion, improving and optimizing traffic flow, and providing in-vehicle entertainment services, etc. [5, 17]. The automotive industry acknowledges the need for an ITS where vehicles are connected to each other through wireless networks [18]. Proper driving is one of the most important factors of traffic safety, so there is a clear need to make it safer. The early warnings of upcoming dangers (e.g., bottlenecks and accidents) would be highly useful for drivers. For this purpose, new kinds of information technology, called vehicular ad hoc network (VANET) and internet-of-vehicles (IoV), are being developed. VANETs are a subset of mobile ad hoc networks (MANETs) in which communication nodes are mainly vehicles. IoV is an extended application of the internet of things (IoT) in intelligent transportation systems. A vehicle can be a sensor platform; it can receive information from the environment or from the other vehicles, and using this information safe navigation, pollution control, and traffic management can be facilitated. The IoV consists of vehicles that communicate with each other as well as with hand-held devices carried by pedestrians, road-side units (RSUs), and public networks. As such, this type of network should deal with a huge number of highly mobile nodes in roads. In VANETs or IoVs, vehicles can communicate with each other through communication network). They can also communicate with an infrastructure (vehicle-to-infrastructure [V2I]) to get some service. This infrastructure is assumed to be located along the roads.

Communication architecture of IoV

Communication in VANETs can be categorized into four types:

1. *In-vehicle communication* is necessary and important in smart vehicle applications. An in-vehicle communication system is used to detect a vehicle's performance and a driver's physical condition, such as fatigue and drowsiness, which are critical for driver and public safety.

2. *V2Vcommunication* provides an information exchange facility for drivers to share information and warning messages, which helps for smoother traveling.

3. *Vehicle-to-road infrastructure (V2I)* communication is another useful tool in the field of VANETs or smart vehicular applications. V2I communication enables the spreading of real-time traffic or weather updates for drivers and provides environmental sensing and monitoring. V2I encompasses the communication between the RSUs, a fixed infrastructure, and a vehicle. Each node in VANET or IoV is equipped with two types of units: anon-board unit (OBU) and other application units(AU). OBU has the communicational capability, whereas AU executes the program making OBU's communicational capabilities. An RSU can be attached to the infrastructure network, which is connected to the Internet.

4. *Vehicle-to-broadband cloud (V2B)* communication provides a way to communicate via wireless broadband mechanisms (e.g.,3Gand 4G). The broadband cloud includes more traffic information and monitoring data, and this information, along with communication, is very useful for active driver assistance and vehicle tracking.

The information for each communication type must be secured. For information security in VANETs or IoV, we move toward the field of cryptography. In this field we have focused on data encryption and decryption. Here, we discuss different encryption techniques, such as data encryption standard (DES), advanced encryption standard (AES), and RC5, and we also discuss the different encryption keys and their robustness. Symmetric encryption or single-key encryption was the only type of encryption in use prior to the development of public key encryption in the 1970s.A symmetric encryption scheme has five parts: Plaintext, encryption algorithm, secret key, cipher text, and decryption algorithm [1, 3].

For the secure use of symmetric encryption, we should focus on two requirements: Strong encryption algorithm and secret key (where the sender and receiver must have copies of this key in a secure fashion). The two basic building blocks of all encryption algorithms are substitution and transposition. There are two types of algorithms: Stream cipher and block cipher. And there are four types of algorithm modes: Electronic codebook (ECB), cipher block chaining (CBC), cipher feedback (CFB), and output feedback (OFB). Symmetric encryption algorithms include DES, triple DES (3DES), international data encryption algorithm (IDEA), blowfish, and AES. Symmetric key algorithm is also known as private key algorithm.

DES [2] uses a block cipher structure. It encrypts data in blocks of size 64 bits each; 64 bits of plain text is treated as the input to DES, which produces 64 bits of cipher text. Initially, the key length is 64 bits. Before starting the DES process, every eight bits of the key are discarded to produce a 56-bit key. DES is based on two main concepts of cryptography: Confusion and diffusion. DES consists of 16 rounds, and each round performs the step of substitution and transposition. DES results in a permutation among the 2^{64} possible arrangements of 64 bits, each of which may be either 0 or 1. Each block of 64 bits is divided into two blocks of 32 bits each, a left half block L and right half R. The DES algorithm turns 64-bit messages block M into a 64-bit cipher block C. If each 64-bit block is encrypted individually, then the mode of encryption is called ECB. There are two other modes of DES encryption, namely CBC and CFB, which make each cipher block dependent on all the previous message blocks through an initial XOR operation.

Double DES [2, 3] is simple to understand, and it performs twice what DES does only once. Double DES uses two keys, say key1 and key2. It first performs as DES on the original plain text using key1to get the encrypted text. It again performs as DES on the encrypted text by using the other key (that is, key2). The final output is the encryption of the encrypted text.

RC5 [2, 3] is a symmetric key block encryption algorithm developed by Ron Rivest. The main theme of RC5 is that it is fast because it uses only primitive operations (XOR, shift, etc.). It allows for a variable number of rounds and a variable bit-size key to add to the flexibility. RC5 uses less memory for execution. In RC5, the word size, number of rounds, and number of octets of the key all may be variable in length, but those values remain the same for a particular execution of the cryptographic algorithm. The output of RC5 is the cipher text, which is the same size as the input plain text.

Recurrence relation in random number generation

A linear recurrence relation is an equation that relates a term in a sequence to previous terms using recursion. Recurrence relation is of two types: Linear recurrence relation and linear non-homogeneous recurrence relation (discussed below) [21].

1. Linear recurrence relation: A linear homogenous recurrence relation of degree k with constant coefficients is a recurrence relation of the form:

   $ka_n = c_1a_{n-1} + c_2a_{n-2} + \cdots + c_ka_{n-k}$ where $c_1, c_2, \ldots, c_k$ are real numbers, and $c_k \neq 0$. a_n is expressed in terms of the previous k terms of the sequence.

   Proposition: Let $ka_n = c_1a_{n-1} + c_2a_{n-2} + \cdots + c_ka_{n-k}$ be a linear homogeneous recurrence.

   - Assume the sequence a_n satisfies the recurrence.
   - Assume the sequence a_n' also satisfies the recurrence.
   - So, $b_n = a_n + a_n'$ and $d_n = \alpha a_n$ are also sequences that satisfy the recurrence. (α is any const.)

2. Linear non-homogeneous recurrence: A linear non-homogenous recurrence relation with constant coefficients is a recurrence relation of the form:

   $a_n = c_1a_{n-1} + c_2a_{n-2} + \cdots + c_ka_{n-k} + f(n)$ where where $c_1, c_2, \ldots, c_k$ are real numbers, and $f(n)$ is a function depending only on n. The recurrence relation $a_n = c_1a_{n-1} + c_2a_{n-2} + \ldots + c_ka_{n-k}$ is called the associated homogeneous recurrence relation.

   Proposition:

   - Let $a_n = c_1a_{n-1} + c_2a_{n-2} + \ldots + c_ka_{n-k} + f(n)$ be a linear non-homogeneous recurrence.
   - Assume the sequence $\{b_n\}$ satisfies the recurrence.

- The sequence $\{a_n\}$ satisfies the non-homogeneous recurrence if and only if:

$h_n = a_n - b_n$ is also a sequence that satisfies the associated homogeneous recurrence.

In our article, we use linear non-homogeneous recurrence relations for generating random numbers. This is another concept to generate random numbers without using rand (). Any recurrence relation can be used in random number generation. For example, here we use the recurrence $a[i] = c[0]*a[i-1] + c[1]*a[i-2] + c[2]*a[i-3] + \ldots + c[m]*a[i-m+1] + pow(2,i)$.

The initial condition and co-efficient values are given at first. According to these conditions and values, the random numbers are generated.

7.2 Literature Survey

In recent years, enormous amounts of research work have been done on the security issues of IoV, which enhanced the performance of services by introducing new protocols. In this section, we will explain some existing methods which have been used to enhance the security of VANETs or IoVs.

Research work on confidentiality

Sun et al. [11] introduced a new security system where a shared key is used for data encryption. This technique is used to protect the confidentiality of sensitive information or data. The aim of this proposed technique is to ensure the confidential information of the registered vehicles and the tracking of vehicles in a legitimate way. It is a fact that the confidentiality of messages is crucial where vehicles get useful data from the internet, Transportation Authority (TA), and RSUs.

Lu et al. [12] introduced a dynamic privacy-preserving key management technique referred to as DIKE, which is used to improve the confidentiality of data in location-based services (LBSs) in the VANET system. To protect the eavesdropping attack, confidentiality must be well maintained. In this method, if a user does not engage in the current VANETsystem, then the user cannot have access to the current LBS content. To gain the confidentiality in an LBS session, a secure session key is shared among all vehicle users who are joined, and that session key can be used to encrypt service contents.

Research work on data integrity

Lin and Li [19] introduce an efficient cooperative authentication technique for the VANET system as well as IoV. This technique is used to shorten the authentication overhead on individual vehicles and to reduce the delay. This method uses a token method to control and manage the authentication workload. This technique comprises the large computational algorithm to control the authentication issue.

Lin et al. [19] introduced a GSIS-based method, which is used to develop a secure privacy-preserving protocol based on the group signature and identity-based signature schemes. In case of dispute, the proposed method can also be used to trace each vehicle, but the ID of the sending message needs to be disclosed by the TA.

Research work on non-repudiation

Li et al. [20] introduced a novel framework with conditional privacy preservation and repudiation (ACPN) for VANETs. This method has utilized the concept of public key cryptography (PKC) to obtain non-repudiation of vehicles by ensuring third parties get the real identities of vehicles. Two methods, such as identity-based signature (IBS) and ID-based online/offline signature (IBOOS) schemes, are utilized for the authentication between V2V and vehicle-to-road-side (V2R) units. These methods significantly reduced computational costs, but the handling of managing certificates is very complex due to IBS and IBOOS authentication schemes.

Research work on privacy-preserving authentication

To achieve a high level of reliability and privacy, Choi et al. [21] introduced a new method that produces a high efficiency of privacy by combining symmetric authentication with the short pseudonyms in VANETs. In this method, to generate short-lived pseudonyms, a transportation authority needs to send the different ID and seed values to each vehicle. RSU performs verification for MACs because it can share keys with vehicles.

Rhim et al. [23] introduced an efficient technique for message authentication based on MAC, but this method cannot tackle and secure against the replay attack. As a result, a sophisticated algorithm is needed and this type of scheme was proposed by Taeho et al. [24] by utilizing the improved MAC authentication scheme for the VANET system.

Chim et al. [17] introduced a method which discussed the security and privacy issues of V2V in VANETs. This scheme has used one-way hash functions and secret keys between vehicles and RSUs for data security. Therefore, this methodology can resolve privacy issues which may occur during communication.

Vighnesh et al. [16] introduced a novel sender authentication technique by using hash chaining and authentication codes to authenticate a vehicle for enhancing VANET and IoV security. This method ensures secure communication between the vehicle and RSU, and a master key is used for data encryption.

He and Zhu [18] presented a method which addresses the problem of denial-of-service (DOS) attacks against signature-based authentication. To tackle DOS attacks, the pre-authentication can be done before signature verification. In this scheme, the pre-authentication mechanism is utilized, which takes advantage of using one-way hash chains and a group rekeying technique.

7.3 Attacks on IoVs

7.3.1 IoVs or VANETs

These are vulnerable to many attacks. Different types of attacks are possible in adhoc network environments, especially in the vehicular domain. These adhoc networks are designed in such a way that makes the situation more complicated. The impact of these attacks over the network primarily depends on the intensions of the attackers behind it. In this case, solutions are proposed by considering some possible attacks to the system once it is implemented.

7.3.2 Impersonation Attack

In an impersonation attack, the attacker represents themselves as an authorized node. These attacks can have the objective of either disturbing the network or gaining access to network privileges. These attacks are possible through identity theft or false attribute possession.

7.3.3 Sybil Attack

In this type of attack, an attacker uses different identities at the same time, i.e., in the VANET, the vehicle announces its various positions at the same time or at frequent intervals of time. It possibly creates confusion and security risks in the network. Sybil attacks harm network topology and can cause bandwidth consumption.

7.3.4 Eavesdropping

This is a threat to confidentiality in adhoc networks. In this attack, the attacker accesses confidential and sensitive data from the network. These attacks fall in the category of passive attacks where attackers silently sense the channel and get the information, and further use that information for their own benefit.

7.3.5 Routing attacks

There are different types of routing attacks, such as grayhole attacks, wormhole attacks, and black hole attack.

7.3.6 Repudiation

The main threat in repudiation is denial or attempt to deny the message by a node involved in the communication system. In this attack, two or more entities have a common identity; hence it is easy to be indistinguishable and they can be repudiated.

7.3.7 Man-in-the-middle attack

This attack occurs in the middle of V2V or V2I communication channel. The attacker can gain access and control the entire V2V or V2I communication, but the communication entities think that they can communicate with each other directly in private.

7.3.8 Symmetric key encryption

In symmetric key encryption we can transmit huge amounts of data between the sender and receiver effectively. But the whole encryption is done using a private key (symmetric key). If this private key is revealed to attackers, then overall communication is under threat. The existing symmetric key encryption algorithm does not change their key/keys with respect to time. So the use of a single fixed key or multiple fixed keys is a problem in the encryption process.

7.4 Solution Domain and Objectives

Our proposed technique provides an extra key called a session key [4] for security purposes. This session key is generated by using a random character matrix using recurrence relation, fuzzy logic, and the symmetric key. Since this session key may change from time to time, the use of both a session key and symmetric key in encryption provides the extra robustness in our technique. Here we deduce a novel technique by using both symmetric key and session key for authentication proof. Thus, the use of a session key with a symmetric key and an authentication cum encryption provides the added flavor to our proposed technique.

7.4.1 Symmetric key generation

In this article, the symmetric key is generated by using the vehicle number, which is produced by the government or a trusted agency, and the driving license number of the driver. Thus, the symmetric key is fixed for a fixed driver with his/her fixed vehicle and when the vehicle or driver changes, the symmetric key may change. Here, the vehicle number is open to all, but the driver's license number is not. So at first, a registration process is necessary with the license number and vehicle number. The advantage of this scheme is that without a license, not just anyone can drive any vehicle.

7.5 Methodology

Our proposed technique is composed of four parts: Session key generation, encryption with symmetric key and session key, authentication check and session key transpired, and decryption. The summary of our scheme is described by a compact algorithm, given below.

```
Algorithm:
Input: Plain text, symmetric key.
Output: Encrypted file with header and tail
------------------------------------------------------------------------
Method:
1. Call MGA ()  // matrix is generated to create "n"
                 number of key populations from the
                 symmetric key and for random number
                 generation linear recurrence relation is
                 used.
2. Call SKG ( )  // generate session key using "n"
                 number of key populations and fuzzy
                 logic [6, 7].
3. Call EP ( )  // encryption process i.e., cipher file
                 is generated using two keys.
4. Call Create_Header_Tail ( )  // header and tail
                                structure is created
                                using two keys with XOR
                                operation.
5. Call Concate(header, encrypted file, tail)  // total
                          structure is created, and it is ready
                          for transmission over the network.
6. Call AuthenticationCheck ( )  // check authentication
                                using two keys and also
                                generate session key
                                using the symmetric key.
7.Call DrypPhase ( )// plain text is generated.
------------------------------------------------------------------------
```

The above methods in the algorithm are described below in detail.

Session key generation phase

The session key generation process is divided into two parts. First is the predefined matrix generation using recurrence relation, and second is the session key generation from the symmetric key using fuzzy logic. The predefined matrix is a square matrix, and the number of columns is half the size of the symmetric key. If the key size is "n" byte then the number of rows and columns of the matrix is n/2.

Algorithm1: Matrix generation algorithm (MGA)

```
Input: Randarr[m]: character array.
Output: A square matrix (kmatrix[m][m]).
Method:
1.Set m, i and j as integer.
2. m= half(symmetric key size).
3. kmatrix[m][m]={0}
4. for i=0 to m
5.    randarr[i]= get_randomchar(); {/* get_randomchar()
generates random character using recurrence relation.*/}
6. end for
7. for i=0 to m
8.    for j=0 to m
9.       kmatrix [i][j]=randarr[j]
10.    end for
11. randarr[i]←rightShft(randarr[i])
12. end for
13.End.
```

Algorithm2: Session key generation (SKG)

```
Input: Symmetric key and kmatrix[x][y].
Output: Session key (SK[n]).
Method:
1. Set i, j, row, col,m fval as integer.
2. Set m= length (symmetric key), tmp[m/2][m/2]= {0},
   SYK[m] = symmetric key and keyarr[m/2],SK[n] as
             character array.
3. row←get_row( kmatrix[m/2][m/2])& col← get_colmn
                              (kmatrix[m/2][m/2])
                                    {/*row=col=m/2*/}
{/* step 4 to step 8 describes key population*/}
4. for i=0 to row do
5.    for j=0 to col do
6.       tmp[i][j]←bitwise_XOROP( SYK[j],kmatrix[i][j])
7.    end for
8. end for
{/* following part of algorithm find the fittest key
among m number of key population using fuzzy logic.*/}
9. Set fval=0
10. for i=0 to row do
11.    fval += bit_Difference (SYK [col], tmp[i][col])
12.      if ((col/fval)<0.04) then
13.        SK[n]←tmp[i][col]
```

```
14.        Keyarr[col]=kmatrix[i][col]
15.      end if
16. end for
17. End.
```

Encryption phase with symmetric key and session key:

Now in our proposed technique, encryption is done by using a symmetric key and session key. The encryption with a session key provides additional protection. In both cases, an XOR [9] operation is executed. The encryption algorithm is given below.

Algorithm3: Encryption process (EP)

```
Input: Plain text, symmetric key, and session key.
Output: Encrypted file.
Method:
1.Set file_Plain as plain text file and file_Cipher as
  cipher text file.
2.Set file_Output as temporary file.
3.if ( !eof ) then
4.file_Output= bit_XOROP ( file_Plain , session key)
5.file_Cipher= bit_XOROP ( file_Output , symmetric key)
6.end if
7.end for
8.End.
```

After encryption with two keys, we create a format with a header, cipher text, and tail [13] using the function *Concate()*. The result of this function is the compact form of text which is ready for transmission to the receiver end. We use the tail part to check authentication and the header part for session key generation at the recipient's end. Now the header and tail structure is created using the following algorithm.

Header and tail creation:

Algorithm4: Create_Header_Tail ( )

```
Input: Symmetric key and session key (SK[m]).
Output: Header and tail.
Method:
1. Set F_half, L_half and diagEl as character arrays.
2. Set key_Mat [][] as 2D character array.
3. F_half← first half of symmetric key, and L_half←
   lasthalf of symmetric key.
4. set m= ascii valueOf(1st character of symmetric key)
```

```
5. Header← ((keyarr[] XOR L_half))<< (m mod length
   (symmetricKey)).
6. Key_Mat← Call create_matrix (F_half, session key)
7. ColmnEl← get_2ndColmn(key_Mat)
8. Tail← bit_XOROP (ColmnEl, L_half)// diagEl is XORed
   withL_half, bit by bit up to the last bit of L_half.
9. End.
```

If the symmetric key is 16 byte, the session key is 8 byte, and the total structure is given below which is created by calling the function *Concate()* which is given in the main algorithm.

Header (8byte)	Encrypted file	Tail (8 byte)

Decryption phase

The decryption phase occurred at the recipient's end. First of all, the header section, encrypted file, and tail section are separated using the symmetric key. Here we call it the*Create_Header_Tail ()* function so that we can reveal the session key using the symmetric key from the header section, and we can check the authentication from the tail part using the function *AuthenticationCheck()*. If the authentication phase shows a green signal then the plain text is generated from the encrypted file (decryption) using both the session key and symmetric key in the reverse process of the encryption phase.

7.6 Results and Discussions

The encryption–decryption process is implemented in the TURBO C interface in a PC of Intel Core i3 processor and 2GB RAM. In this section, simulation results of the proposed scheme are presented. In our experiments, several sizes of files are used as plain text.

Different types of attacks

Different types of attacks exist to recover the key in use rather than simply to recover the plain text. Different types of attack analysis (brute-force attack [11, 12]), experiments and their results and comparisons are discussed below.

Key space analysis

The size of the key space is the total number of different keys in the encryption process. The brute-force attack is impractical in such crypto systems where the key space is large. Now we consider a general case where the secret key is "k" bits. There are two keys are in our proposed scheme: the first is a symmetric key with the size "k" bits, and the second is a session key whose size is k/2 bits. Now for the symmetric key, the key space is 2^k, and for session key, the key space is $2^{k/2}$ and total key space is $2^{3k/2}$. Using this large key space, we have discussed brute-force attack [10, 22].

Brute-force attack:

> A good encryption technique satisfies the requirements of avoiding a brute-force attack. In this attack, the attacker tries to translate the cipher text into plain text using every possible key. On average, half of all possible keys must be attempted to achieve success. In most networking systems, algorithms are known to all, so in this case, a brute-force attack will be impossible if the algorithm uses a large number of keys. At present, the fastest super computer is Tianhe-2, having a speed of 33.86 petaflops, i.e., 33.86×10^{15} floating-point operations per second. Let us consider each trial requires 2,000 FLOPS to complete one check. So the number of trials completed per second is 16.93×10^{12}. The number of seconds in a year is 365*24*60*60=3153600 seconds.

From the above discussion it is seen that the time for break the keys is $2^{3k/2}$/ $(16.93 \times 10^{12} \times 3153600)$ = Y. So if k increases then Y increases. Figure 7.1 and Table 7.1 show the average time required for an exhaustive key search [3].

Observations

From Table 7.1 it is seen that with respect to the number of trials, our proposed technique provides better results than any standard algorithms (like DES, Triple DES, AES, etc.) with the same key size. The x-axis of the graph represents the key size in bits. Figure 7.1 and Table 7.1 also show that our proposed technique provides better results for decryption than any standard

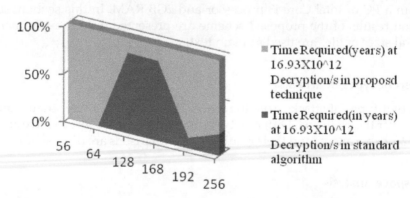

FIGURE 7.1
3D graph of an exhaustive key search.

TABLE 7.1

Exhaustive Key Searches

Symmetric key size (k bits)	No. of trials in standard algorithms(2^k)	Time required (years) at 16.93×10^{12} Decryption/s in standard algorithms	No. of trials in our proposed technique ($2^{(3k/2)}$)	Time required (years) at 16.93×10^{12} Decryption/s in the proposed technique
56	2^{56}	0.001349	2^{84}	362289
64	2^{64}	0.34550	2^{96}	1483938
128	2^{128}	6.3734×10^{18}	2^{192}	1.1775×10^{38}
168	2^{168}	7.0077×10^{30}	2^{252}	1.3555×10^{56}
192	2^{192}	1.1756×10^{38}	2^{288}	9.3148×10^{66}
256	2^{256}	2.1687×10^{57}	2^{384}	7.3799×10^{95}

algorithms (like DES, Triple DES, AES, etc.) with fixed decryption rates. So, it is difficult for an attacker to decrypt any cipher text using an assumed key. Thus, the overall results of our technique are good with respect to any standard algorithms in a brute-force attack.

Randomness test of session key

In our technique, the session key is generated from a symmetric key using fuzzy logic. To test the randomness of the session key, we use some standard techniques such as a frequency test [14] and entropy [15].

Frequency test

The frequency test is the most basic test for randomness checking. The purpose of this test is to determine whether the number of ones and zeros in a sequence is approximately the same as would be expected for a truly random sequence. Figure 7.2 and Table 7.2 represent the details.

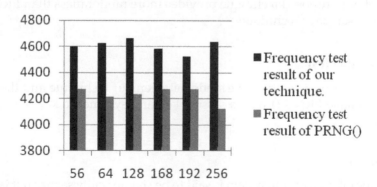

FIGURE 7.2
Graph of frequency test of the above Table 7.2.

TABLE 7.2

Frequency Test Results

Symmetric key size (bits)	Frequency test result of our technique	Frequency test result of PRNG()
56	4598.255	4272.771
64	4625.148	4210.445
128	4661.584	4235.617
168	4579.129	4270.673
192	4519.966	4270.756
256	4633.421	4123.350

Observations

NIST SP 800-22 specifies that the randomness test must follow three characteristics: Uniformity, scalability, and consistency.

In the case of uniformity and scalability, the occurrence of zero or one is equally likely; that is, the probability of occurrence of zero or one is 0.5. The above table of frequency test result shows uniformity and scalability of our technique.

In case of consistency, we can say that the seed value from which we can generate the session key is a symmetric key. For cryptographic applications, the symmetric key must be secure. The session key is generated by using a random key matrix and a symmetric key. If the key matrix is unknown or changes from time to time and if the symmetric key is secured, then the next output bit in the sequence should be unpredictable in spite of any knowledge of previous bits in the sequence.

It should not be feasible to determine the symmetric key from the knowledge of any generated values. There is no correlation between the symmetric key and generated values. Thus, our technique proves the forward and backward unpredictability. Furthermore, from Table 7.2 and Figure 7.2 it is seen that our proposed technique provides more randomness than PRNG (), which is a standard technique.

Entropy value test

Here we describe a comparative study between our technique and the standard technique PRNG () with a session key and symmetric key (Figure 7.3) (Table 7.3).

Observations

In cryptography, a cryptosystem is said to be semantically secure if it is computationally impossible for an attacker to extract any information about the plain text from the cipher text. Entropy can be defined as randomness or

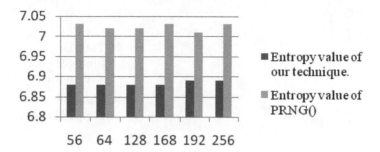

FIGURE 7.3
2D graph of the entropy value of the above Table 7.3.

TABLE 7.3

Entropy Values

Symmetric key size (bits)	Entropy value of our technique	Entropy value of PRNG()
56	6.88	7.03
64	6.88	7.02
128	6.88	7.02
168	6.88	7.03
192	6.89	7.01
256	6.89	7.03

unpredictability of information contained in a message. This randomness breaks the structure of the plain text. Entropic security in encryption is similar to semantic security when data have a highly entropic distribution. Plain text entropy value is zero. From the comparative study of entropy values between our technique and PRNG (), it is seen that the entropy value of our technique is near to the result of PRNG (). The x-axis (Table 7.3 and Figure 7.3) represents the key length. Thus, from the definition of entropic security we can say that it is difficult to predict plain text from cipher text if we use our technique to generate a session key. The use of this session key and symmetric key in encryption provides robustness.

Comparative study on avalanche effect

Here, a comparative study on avalanche effect [3, 16] with a fixed key is described below with a table and graph(Figure 7.4) (Table 7.4).

Observations

A desirable property of any encryption algorithm is that a small change in either the plaintext or the key should produce a significant change in the cipher text. In particular, one-bit change in the plaintext or one bit in the key

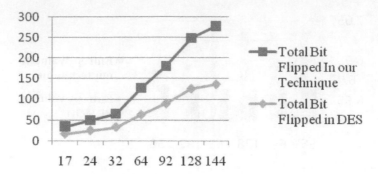

FIGURE 7.4
Graph of avalanche effect with fixed key for the above Table 7.4.

TABLE 7.4

Avalanche Effect with Fixed Key

Text size (byte)	Total bit flipped in DES	Total bit flipped in our technique
17	17	17
24	25	24
32	33	32
64	63	64
92	90	90
128	126	123
144	137	141

should produce a change in many bits of the cipher text. Thus, avalanche [3] quantifies the effect on the cipher text when there is one-bit change in plain text. An encryption algorithm that does not provide the avalanche effect can lead to an easy statistical analysis if the change of one bit from the input leads to the change of only one bit of the output. Figure 7.4 and Table 7.4 provide a comparative study between DES and our technique. In the graph, the x-axis represents the text size. This study tells that the total number of bit flipped in our encryption technique is more than DES. Here we use a fixed size key. Thus, our technique (using a fixed key) provides better results than any standard algorithm (like DES). So, our proposed scheme satisfies the desirable property for an encryption algorithm.

Robustness of our encryption technique

Cryptanalysis is the study of cipher text, ciphers, and cryptosystems. The aim of cryptanalysis is to understand how they work and finding and improving the techniques for defeating or weakening them. There are

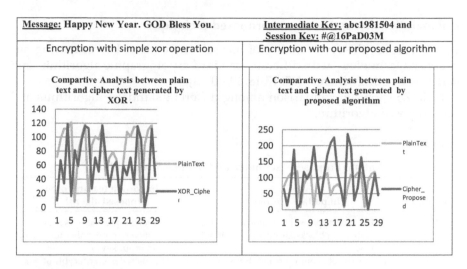

FIGURE 7.5
Analysis of an encryption technique.

different types of cryptanalysis attacks, such as a cipher text-only attack, chosen plain text attack, known plaintext attack, and chosen cipher text attack. Breaking an encryption algorithm is basically the finding of the key to access the encrypted data in plain text. For symmetric key encryption, breaking the algorithm usually means trying to determine the key used to encrypt the text. For a public key encryption, breaking the algorithm usually means acquiring the shared secret information between two recipients. The robustness of an encryption technique depends on non-linearity in cipher text. In our paper, we use circular left shift operation and a non-linear function to provide non-linearity in cipher text. As a result, our technique is able to protect any type of cryptanalysis attack. The graph in Figure 7.5 shows the robustness of our protection mechanism.

Observations

Non-linearity is the main theme in any encryption technique. From the above Figure 7.5 it is seen that our technique provides more non-linearity in cipher text than a simple XOR operation. If we consider any point (x, y) and (a, b) in the above graph, then any periodic gap does not exist between two points in the graph; i.e., there is no relationship between the two graphs. So it is hard to guess the plain text or encryption key from the cipher text. Thus, our encryption scheme is robust and it may protect any type of cryptanalysis like known plain text attack, chosen cipher text attack, etc. It is seen that our technique satisfies the condition of perfect security because cipher text and plain text are independent.

Comparative discussion of our proposed technique

In this section, the functionality of our scheme is done by comparing our proposed technology with different standard cryptographic algorithms and different existing schemes [15, 17, 18, 19, 20, 21].

Table 7.5 shows a comparison among different standard algorithms and our proposed algorithm.

TABLE 7.5

Comparisons Among Proposed Technique and Standard Algorithms

Algorithms	Important features	Important features of our proposed algorithm
IDEA	i) IDEA encrypts 64-bit plaintext to 64-bit cipher text blocks, using a 128-bit input key. ii) It uses both confusion and diffusion technique. iii) A dominant design concept in IDEA is mixing operations from three different algebraic groups of 2^n elements. iv) The security of IDEA currently seems that it is bounded only by the weaknesses arising from the relatively small (compared to its key length) block length of 64 bits.	i) Our technique encrypts n-bit plaintext to n-bit cipher text, using m-bit input key. ii) The main design concepts of our technique: (a) intermediate key generation using the concept of linear congruence, (b) session key generation using approximation algorithm, and (c)circular left shift is usedto produce non linearity in encryption process.
RC5	i) The RC5 block cipher has a word-oriented architecture for variable word sizes w = 16, 32, or 64 bits. ii) The number of rounds r and the key byte-length b are variable. iii) For encryption, there are three steps in each round: (a) bit-wise XOR operation, (b) circular left shift, and (c) addition with the next sub key.	i) Our proposed scheme is stream cipher based. Here two keys are used for encryption/decryption. ii) Key length is variable. iii) For encryption, there are two steps:(a) bit-wise XOR operation, and (b) circular left shift with a linear function. It provides the number of times CLS occurs.
Blowfish	i) This technique is based on stream cipher. It uses an additional XOR operation for encryption. ii) It has a variable key length up to a maximum of 448 bits long, which ensures security. iii) Blowfish suits applications where the key remains constant for a long time, and it is not suitable for packet switching.	i) Our scheme is also based on steam cipher. It uses XOR and CLS operations to impose more non-linearity in cipher text. ii) It uses double keys, and one of these keys is changeable by nature which provides robustness in our technique. iii) Suitable for packet switching.
DES	i) Linear cryptanalysis provides the most powerful attack on DES to date, where an enormous number of known plain text pairs are feasible.	i) Our algorithm is based on stream cipher with two keys, one of which is a session key, which is changeable in nature. So it protects linear cryptanalysis as well as differential cryptanalysis.

(Continued)

TABLE 7.5 (CONTINUED)

Comparisons Among Proposed Technique and Standard Algorithms

Algorithms	Important features	Important features of our proposed algorithm
	ii) Differential cryptanalysis is one of the most general cryptanalytic tools to date against modern iterated block ciphers, including DES. It is primarily a chosen-plaintext attack. iii) Storage complexity, both linear and differential cryptanalysis, requires only negligible storage. iv) Due to its short key size, the DES algorithm is now considered insecure and should not be used. However, a strengthened version of DES called triple-DES is used.	ii) Our algorithm takes negligible storage for linear and differential cryptanalysis. iii) Our algorithm is secure with respect to key size because we have used two keys with variable length.
Triple-DES with three keys	i) Triple-DES counters the meet-in-the-middle attack by using three stages of encryption with three keys. ii) Tuchman proposed a triple encryption method that uses only two keys. The function follows an encrypt-decrypt-encrypt sequence. $C=E(K_1,D(K_2,E(K_1,P)))$ and $P=D(K_1,E(K_2,D(K_1,C)))$ iii) There is no cryptographic significance in second stage decryption. Its only advantage is that it allows users to decrypt information encrypted by users of single DES. iv) There is no practical cryptanalytic attack on 3DES. This method is an improvement over the chosen plain text approach but requires more effort. This attack is based on the observation that if the value of the first phase encryption and final cipher text is known, then the problem reduces to double DES.	i) This article proposed a method that counters a meet-in-the-middle attack by using two different keys with variable times of CLS. ii) This technique has cryptographic importance in a wireless network; it uses two keys, an intermediate key and session key, which are generated using fuzzy logic. It also provides strong authentication mechanisms. iii) It is very hard to anticipate the two keys if the plaintext-cipher text pair is known.

Significance of authentication

Authentication mechanisms [8] provide proof of identities. The authentication process ensures that the origin of a document is correctly identified, i.e., the document is coming from the right person. In our scheme we use authentication as proof of identity. We know that a symmetric key encryption provides authentication and confidentiality. Here we have also proposed an authentication using symmetric key. Here we use two structures: header and tail. There are complex calculations for header and tail generation. Tail structures are used for authentication checks. In the receiver side, the session key is generated first and then using this session key and symmetric key we can check authentication from the tail part. Thus, our technique protects against fabrication.

7.7 Conclusion

Smart vehicular systems have become an essential platform for information exchange between vehicles, humans, and road-side infrastructures, and this information must be secured and protected from access by an intruder or attacker. This is because the dissemination of incorrect information between vehicles on the road may cause road accidents or vehicle hijackings. Thus, in vehicular networks, trust and security are two main issues. For security we have presented an encryption method based on a symmetric key and session key. This session key is generated from a symmetric key using certain tools such as fuzzy logic and a random matrix. Here, the receiver decrypts the cipher text using his or her symmetric key and session key. Again, trust formation is a serious problem in IoV, and our technique also provides proof of identities which enrich the robustness as well as trust in a vehicular network. Thus, our technique provides two contributions to symmetric key encryption, which is used in IoV. Comparative statistical tests like entropy value and frequency tests between the proposed techniques and standard techniques prove the sturdier of our key. Lastly, an exhaustive key search analysis shows the acceptability of our technique. To the best of our knowledge, our proposed technique is the simplest encryption technique with a symmetric key, and session key with authentication mechanisms. It practically has minimal computational overhead during encryption and decryption.

References

1. A. Das and C. E. Veni Madhavan, *Public-Key Cryptography: Theory and Practice* (Pearson Education, India).
2. D.Boneh and M.Franklin, Identity-Based Encryption from the Weil Pairing, in Kilian, J. (ed.), *CRYPTO2001*. LNCS, vol. 2139 (Springer, Heidelberg, 2001), pp. 213–229.
3. W.Stallings, *Cryptography and Network Security: Principles and Practice*, third edition (Prentice Hall, India 2003).
4. Fuzzy Logic: An Introduction. [Online] Available: http://www.seattlerobotics.org.
5. T. Yamakawa, Electronic circuits dedicated to fuzzy logic controller: Scientia Iranica, Volume 18, Issue 3, June 2011, Pages 528–538.
6. *Fuzzy Sets and Applications: Selected Papers by L. A. Zadeh* (R. R.Yager, et al., eds.) (John Wiley, New York, 1987).
7. Hamzah, A.; Al-Jarrah, M.S.O.; Taqieddin, E. Energy-efficient fuzzy-logic-based clustering technique for hierarchical routing protocols in wireless sensor networks. Sensors 2019, 19, 561.

8. AtulKahate, *Cryptography and Network Security* (Tata McGraw-Hill Publishing Company, New Delhi, 2008).
9. E. T.Oladipupo and O. A.Alade, An Approach to Improve Data Security Using Modified XOR Encryption Algorithm. *International Journal of Core Research in Communication Engineering,* 1(2) (2014),pp. 200–210.
10. D.Stinson, *Cryptography: Theory and Practice,* third edition (Chapman & Hall/ CRC(India), 2006).
11. A.Agrawal, S.Gorbunov, V.Vaikuntanathan, and H.Wee, Functional Encryption: New Perspectives and Lower Bounds, in R.Canetti, and J. A.Garay, (eds.) *CRYPTO 2013, Part II.* LNCS, vol. 8043 (Springer, Heidelberg, 2013), pp. 500–518.
12. J. Buchmann, *Introduction to Cryptography,* second edition (Springer(UK), 2004).
13. S. A.Chaudhry, et al. An Improved and Provably Secure Privacy Preserving Authentication Protocol for SIP. *Peer-to-Peer Networking and Applications*10(1) (2017): pp.1–15.
14. A.Kak, Lecture Notes on Computer and Network Security (Purdue University, 2015). [Online] Available: https://engineering.purdue.edu/kak/compsec/Le ctures.html.
15. B. Zaidan, A. Zaidan, A. Al-Frajat, and H. Jalab, On the Differences Between Hiding Information and Cryptography Techniques: An Overview.*Journal of Applied Sciences*10 (2010): 1650–5.
16. H.Delfs and H.Knebl, *Introduction to Cryptography: Principles and Applications* (Springer(USA), 2002).
17. M.Nidhal, J.Ben-othman, and M.Hamdi, Survey on VANET Security Challenges and Possible Cryptographic Solutions. *Vehicular Communications*1(2) (2014): 53–66.
18. Z.Lu, G.Qu, and Z.Liu, A Survey on Recent Advances in Vehicular Network Security, Trust, and Privacy.*IEEE Transactions on Intelligent Transportation Systems*20(2) (2019): 760–776.
19. P.Papadimitratos and Z. J.Haas, Secure Data Transmission in Mobile Ad Hoc Network. *ACM Workshop on Wireless Security,* San Diego, CA, September 2003.
20. K.Sanzgiri, B.Dahill, B. N.Levine, C.Shields, and E. M.Belding-Royer, A Secure Routing Protocol for Ad Hoc Networks. *Proceeding of IEEE ICNP*2002 (2002): 78–87.
21. Y. C. Hu, A. Perrig, and D. B. Johnson, Ariadne: A Secure On-Demand Routing Protocol for Ad Hoc Networks. *Mobi Com*02 (2002): 23–26.
22. M.Guerrero and N.Asokan, Securing Ad Hoc Routing Protocols.*Proc.1st ACM Workshop on Wireless Security,* 2002, pp. pp. 1–10.
23. DavideCerri and AlessandroGhioni, Securing AODV: The A-SAODV Secure Routing Prototype.*IEEE Communications Magazine,* Feb2008. pp. 120–126.
24. Y. C.Hu, D. B.Johnson, and A.Perrig, SEAD: Secure Efficient Distance Vector Routing for Mobile Wireless Ad Hoc Networks. *Ad Hoc Networks*1 (2003):175–192.

Part 3

Smart Safety Measures and Applications

8

Hybrid Data Structures for Fast Queuing Operations in Vehicular Communication

Raghavendra Pal

CONTENTS

ABSTRACT   Vehicular communication is used not only for safety purposes but also for assistance and infotainment purposes. The priorities of packets need to be assigned because they help vehicles in determining the sequence in which the packets should be transmitted. Safety packets are given the highest priority and infotainment packets are given comparatively low priority. However, these are not the only classifications. Safety and infotainment messages are further divided into subclasses. Furthermore, the priority of a packet also depends upon the characteristics of the vehicle that is transmitting the packet. The speed of vehicle, the distance from the receiver vehicle, and other parameters are involved in determining the priority of the packets in the queue of the vehicle. After the calculation of priorities, packets are arranged in a particular data structure which can be either a linked list or a tree or a combination of both. This chapter discusses the major classifications of the messages in the vehicular communications and the possible data structures which can increase the efficiency of the queue, i.e., decrease enqueue and dequeue times.

8.1 Introduction

Vehicular communication has been an area of research for more than a decade. It aims to provide safety to vehicles by alerting them to any unusual situation. Another area of vehicular communication is to provide infotainment services to the driver, such as traffic information and map downloading. The Institute of Electrical and Electronics Engineers (IEEE) has defined the physical and medium access control layer parameters in a standard named IEEE 802.11p [1] as shown in Table 8.1. A combination of IEEE802.11a [2] and IEEE802.11e [3] form the standards for wireless LANs. The distributed coordination function (DCF) mechanism is taken from IEEE802.11a, which is based upon carrier sense multiple access/collision avoidance (CSMA / CA). The provision of quality of service (QoS) is taken from IEEE802.11e. In the United States of America, the United Nations Federal communication commission (FCC) has allotted a spectrum called dedicated short-range communication [4] for the purpose of vehicular communications. DSRC is a 75 MHz band with 7 channels of 10 MHz each, and 5 MHz is provided as guard band to separate it from nearby bands, as shown in Figure 8.1. The main use of vehicular communication is to provide safety to vehicles on the roads. It does this task by transmitting information that is related to an event that can cause an accident, or by communicating information if an accident has already occurred. After reception of the message, nearby vehicles alert their drivers that there is an unusual situation in the area. The driver can act accordingly after receiving vehicle's alert. Another application of vehicular communication is the transmission of infotainment messages. Infotainment messages are those which are not directly related to accident or safety messages. Examples of infotainment messages are map extraction, traffic information, optimum route information, nearby amenities, etc. These types of messages are less important than the safety messages, so whenever the vehicle has to transmit any one of these messages, it should always transmit safety message first. However, this classification of

TABLE 8.1

Some MAC and PHY Layer Parameters in Vehicular Communications

Parameters	Values
Short interframe space (SIFS)	32 μs
T_{slot}	16 μs
AC	VO, VI, BE, BK
AIFSN	Depends on AC
AIFS	Depends on AIFSN and T_{slot}
CW_{min}	15
CW_{max}	1023
Transfer rate	3–27 Mbps
Bandwidth	10 MHz

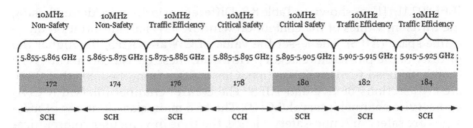

FIGURE 8.1
Dedicated short-range communication band.

messages is not enough. Safety messages can be further divided into subcategories. For example, if there is an accident on road, a safety message should be transmitted immediately. However, if at the same time, there is a packet that contains information about a tree that has fallen on the road, which message should be transmitted first? It is pretty clear in this example that in spite of both being safety messages, the first message is of higher priority since the accident has occurred and the passengers inside the vehicle will be in need of medical assistance. However, there can be confusion in deciding the priority of the packet. For example, the fallen tree in the road is a type of safety message since it can cause an accident. On the other hand, the sudden breaking down of a vehicle is also a safety message since it can cause an accident too. Which message will be transmitted first? There is some confusion since both messages seem equally important. To avoid this confusion, other parameters are also taken into account when deciding the priority of the message. Some of these parameters include the speed of the vehicle transmitting the packet, the size of the packet, the remaining lifetime of the packet, and the distance between transmitter and receiver. This type of priority is called dynamic priority, and the priority based on the type of message is called static priority. Both the priorities are used to determine the priority of a packet.

Whenever a vehicle receives a packet, it stores it in a queue so that it can be transmitted whenever the channel is idle. If another packet is generated at the vehicle or the vehicle receives another packet, this packet is also enqueued in the queue. If the queue size is large, the data structure that is used to store the packet becomes important. It is because the efficient data structure enables the enqueue and dequeue operation quickly. This chapter first discusses the possible classification of packets and their priorities and then the possible data structures which can be used to store the packets.

8.2 Literature Survey

Researchers have been working in the area of vehicular communication for over ten years. IEEE has defined broad priorities in its standard

IEEE802.11p [1], as shown in Table 8.2. Different priorities are defined using the different values of the contention window (CW) and arbitration inter-frame space (AIFS). The lesser the value of CW and AIFS, the higher the priority of the particular category. However, these classifications are very broad and need further elaboration. One can see from the table that there are two sections of CW and AIFS: The control channel interval (CCHI) and service channel interval (SCHI). The two main classifications of messages are safety and non-safety. Hence, the time in vehicular communication is divided into two sections: CCHI SCHI. These interval times are 50 milliseconds each. Similarly, the seven channels in the dedicated short-range communications band are divided into two categories: CCH and SCH. CCH in only one, i.e., there is only one channel dedicated for control purpose. The same channel is also used for the transmission of safety messages. On the other hand, there are six SCHs. These are used in the SCHI for the transmission of messages other than safety messages. CCHI is also used for the transmission of beacons. These are the control packets which allow all the vehicles to receive their neighborhood information. At the beginning of the CCHI, every vehicle transmits its beacon to neighboring vehicles. Since the network in the vehicular communication is ad hoc, beacon transmission is required. Researchers in [5] have developed a new algorithm that ensures the successful transmission of beacons.

The messages in vehicular communication are classified, as shown in Table 8.3. They have divided the priorities into four classes. The message generation type is also defined, i.e., whether the message is event driven or periodic. The dedicated short-range standard for the United States can be found in [6]. Authors in [7–13] have proposed algorithms which transmit either the safety message or non-safety message or both effectively.

All the packets are stored in a data structure in the vehicle. Whenever a packet is received by the vehicle or a packet is generated by the vehicle, it first stores the packet in that data structure. After that, when it finds the channel idle, the packet of highest priority is transmitted first. When a packet is received by the vehicle, the packet needs to go all the way to its place in the queue according to its priority. This may take some time when the queue

TABLE 8.2

Different Access Categories

AC No.	Access category	CCH interval				SCH interval			
		CW_{min}	CW_{max}	AIFSN	AIFS (µs)	CW_{min}	CW_{max}	AIFSN	AIFS (µs)
0	Background traffic (BK)	15	1,023	9	144	15	1,023	7	112
1	Best effort (BE)	7	15	6	96	15	1,023	3	48
2	Video (VI)	3	7	3	48	7	15	2	32
3	Voice (VO)	3	7	2	32	3	7	2	32

TABLE 8.3

Message Classifications in Vehicular Communication

Applications	Priority type	Priority class	Network traffic	Message range (m)
Accident information	Safety	Class 1	Event driven	300
Crossroad warning	Safety	Class 1	Event driven	300
Sudden acceleration or deceleration	Safety	Class 1	Event driven	<50
Accident-prone area warning	Safety	Class 2	Event driven	300
Cooperative accident avoidance	Safety	Class 2	Periodic	300
Toll plaza information	Non-safety	Class 3	Event driven	300
Parking information	Non-safety	Class 3	Event driven	300
Entertainment messages	Infotainment	Class 4	Event driven	300
Amenities information	Infotainment	Class 4	Event driven	0–90

is implemented in the hardware. Further, the dynamic priority of a vehicle changes with time, so a packet that currently has low-dynamic priority may attain high-dynamic priority after some time. This also causes the frequent rescheduling of a queue. So there is a need for a data structure that allows fast searching and enqueue and dequeue operations. Authors in [7] have proposed a data structure that is based upon the static and dynamic priorities of a message and reduces the time for enqueue and dequeue operations.

8.3 Message Classifications and Priority Assignment in Vehicular Communications

The major classifications of messages in vehicular communications have been provided in Table 8.3. However, the classifications can further be divided into subclasses. For example, accident information message can be of many types, i.e., place and time of accident, extreme weather conditions, vehicles out of control, and fuel tank leakage warning. Similarly, amenities information can be of many types. The possible classifications of messages are shown in Table 8.4. However, there can be many other messages. The static priority assigned to each message is solely based upon the type of the message. The priority of each message type may or may not be different; however, in most cases it is different. In case the static priority is not different for two or more message types, the dynamic priority is used to find out that which message is to be transmitted first. Dynamic priority is also used to identify the priority of two or more of the same type messages.

Table 8.4 defines static priority only. To calculate dynamic priority, some parameters are taken into account. These parameters are the speed of the vehicle, the distance between the transmitting and receiving vehicle, message size, deadline of the message, etc. These parameters are combined to

TABLE 8.4

Further Classification of Message Types in Vehicular Communications

Applications	Message type	Priority type	Priority class
Accident information	Vehicle accident occurred	Safety	Class 1
	Extreme weather conditions, hurricanes etc.		
	Fuel tank leakage warning		
	Unconscious driver		
	Break failure of a vehicle		
Crossroad warning	Intersection ahead	Safety	Class 1
	Sharp curve ahead		
Sudden acceleration or deceleration	Sudden breaking	Safety	Class 1
Accident-prone area warning	Tree fallen on the road	Safety	Class 2
	Road pits ahead		
	Construction in progress		
	Slippery road		
Cooperative accident avoidance	Diversion ahead	Safety	Class 2
	Rash driving		
Toll plaza information	Electronic toll collection	Non-safety	Class 3
Parking information	Automatic parking	Non-safety	Class 3
Entertainment messages	Infotainment service access	Infotainment	Class 4
Amenities information	Map downloading	Infotainment	Class 4
	Traffic information		
	Best route finding		
	Car services		
	Fuel station information		

form an equation to calculate the dynamic priority as in [14]. This equation is intelligently defined. For example, let us assume a simple equation to calculate the dynamic priority of a message:

$$\text{Dynamic Priority} = \frac{\text{distance between sender and receiver} + \text{speed of the vehicle}}{\text{message size} + \text{data rate}}$$

Looking at the numerator part of the equation, as the distance between sender and receiver increases, the chances of them getting out of each other's range are high. Hence, as the value of the distance is increased the dynamic priority should be increased, meaning that distance and dynamic priority are directly proportional to each other. Similarly, if the message size is less, it will take comparatively less time to transmit the message, hence this message can be transmitted with high priority. For this reason, dynamic priority and message size are inversely proportional to each other. The explanation for other two variables is similar.

8.4 Quick Enqueue and Dequeue Operations

As already mentioned, the packets in the queue are required to be frequently rearranged due to two reasons. First is the reception and generation of messages and the second is a change in the dynamic priority of the message due to its deadline. This frequent rescheduling of queues can cause additional delay to the messages. Researchers have paid little attention towards this area. Authors in [7] have proposed a hybrid data structure consisting of a linked list and several red–black search trees. The linked list contains the packets with different static priorities. Each node on the linked list acts as the root node of the red–black search tree. Each tree contains packets of the same static priority and different dynamic priorities, as shown in Figure 8.2.

Whenever a vehicle receives a packet for forwarding, it stores the packet in the hybrid tree. The linked list is first checked for similar static priority nodes in comparison to the packet. When it searches it, the correct position in the corresponding red–black search tree is identified according to its dynamic priority. The authors have compared the enqueue and dequeue times of the packets with the linked list. Figures 8.3 and 8.4 show the enqueue and dequeue times, respectively, for the linked list and hybrid tree.

As the number of pre-existing packets in the queue increases, the enqueue and dequeue times increase as well. However, the performance of the hybrid tree is much better in comparison to the linked list. It shows the effectiveness of the data structure in reducing the queuing delay.

This work motivates us to explore some other forms of data structures. They have combined the linked list and red–black search tree to form a hybrid data structure. Some other combinations can also be used for the reduction of enqueue and dequeue times. The first of these combinations is

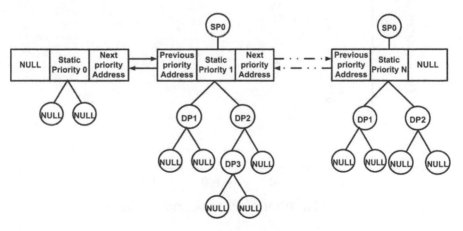

FIGURE 8.2
Hybrid tree for fast enqueue and dequeue operations.

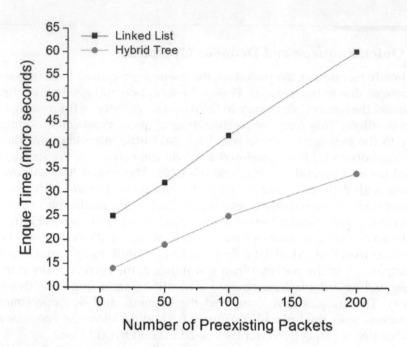

FIGURE 8.3
Enqueue time of linked list and hybrid tree.

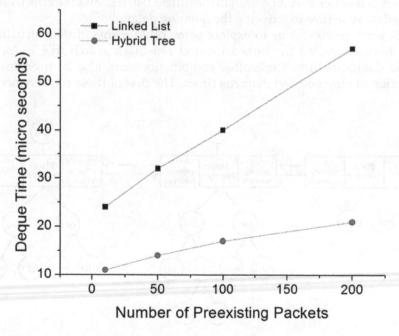

FIGURE 8.4
Dequeue time of linked list and hybrid tree.

a hybrid of two linked lists, as shown in Figure 8.5. The first linked list contains the packets of different static priorities. Every other linked list contains the packets of the same static priorities but different dynamic priorities. If a packet is received by the vehicle, it will first search for the node in the first linked list which has a similar static priority as that of the packet. Then it will search the corresponding linked list which contains all the packets of the same static priority. It will go all the way to the position where the dynamic priority of the packet is lesser than the next packet and higher than the previous packet. This data structure will cause additional delays in comparison to the one proposed in [7].

The next possible data structure can be the combination of the linked list and the binary search tree and will be similar to the one shown in the [7]. The only difference is that it uses the binary search tree instead of the red–black search tree, as shown in Figure 8.6. Since the red–black search tree is more efficient than the binary search tree, this data structure will have higher delays in comparison to [7].

Other data structures which may perform well in comparison to the one proposed in the [7] can be ones which use the binary search tree for the storage of both priority types. The different static priority messages are arranged

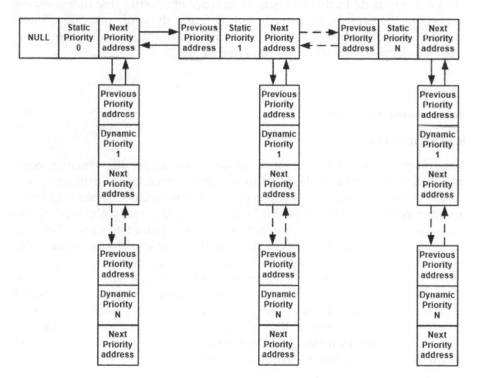

FIGURE 8.5
Hybrid data structure consisting of linked lists for static as well as dynamic priorities.

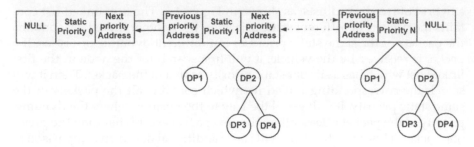

FIGURE 8.6
Data structure consisting of linked list and binary search trees.

in a binary search tree and each node of this binary search tree acts as the root for another tree which has packets of the same static priorities but different dynamic priorities. This data structure may or may not perform well in comparison to [7]. It requires an analysis using any simulation tool or hardware. The most efficient data structure that has the highest chances of performing well in comparison to [7] is a combination of red–black search trees. One red–black search tree contains the packets of different static priorities. Every node in this tree acts as the root of another tree that contains packets of the same static priority but different dynamic priorities. There may be other data structures which can be used in the vehicular communication. Extensive research is required in this area.

8.5 Summary

This chapter presented the classification of messages in vehicular communication. It highlights the ways in which the priorities of different packets are defined. It talks about two types of priorities: One is a static priority, and another is a dynamic priority. Static priority is directly obtained by the message type. To calculate dynamic priority, some parameters are taken into account. These parameters are the speed of the vehicle, distance between the transmitting and receiving vehicle, message size, deadline of the message, etc. Further, it talks about the data structure in which the packets are scheduled. The frequent rescheduling of queues makes it necessary to develop an efficient data structure which allows fast enqueue and dequeue operations. Otherwise these operations can cause additional delays to packet transmission. The possible combination of various data structures is also described. Researchers can take it into account when designing more efficient data structures.

References

1. "IEEE Standard for Information Technology – Local and metropolitan area networks – Specific requirements--Part 11: Wireless LAN medium access control (MAC) and physical layer (PHY) specifications amendment 6: Wireless access in vehicular environments," in IEEE Std 802.11p-2010, pp.1–51, July 2010, doi: 10.1109/IEEESTD.2010.5514475.
2. H. Lee, J. Lee, S. Kim, and K. Cho, "Implementation of IEEE 802.11a wireless LAN," *2008 Third International Conference on Convergence and Hybrid Information Technology*, Busan, 2008, pp. 291–296.
3. A. Grilo and M. Nunes, "Performance evaluation of IEEE 802.11e," *The 13th IEEE International Symposium on Personal, Indoor and Mobile Radio Communications*, Pavilhao Altantico, Lisboa, Portugal, 2002, pp. 511–517, vol. 1.
4. N. Gupta, A. Prakash, and R. Tripathi, "Medium access control protocols for safety applications in vehicular ad-hoc network: A classification and comprehensive survey," *Vehicular Communications*, 2(4), (2015): 223–237. doi: 10.1016/j.vehcom.2015.10.001
5. Raghavendra Pal, Nishu Gupta, Arun Prakash, and Rajeev Tripathi, "Adaptive mobility and range based clustering dependent MAC protocol for vehicular ad-hoc networks," *Wireless Personal Communications*, Springer, 98 (2018): 1155–1170. doi: 10.1007/s11277-017-4913-9
6. B. J. Kenny, "Dedicated short-range communications (DSRC) standards in the United States," *Proceedings of the IEEE*, 99(7) (2011): pp. 1162–1182.
7. R. Pal, A. Prakash, R. Tripathi, and K. Naik, "Scheduling algorithm based on preemptive priority and hybrid data structure for cognitive radio technology with vehicular ad hoc network," *IET Communications*, 13(20) (2019): 3443–3451.
8. R. Pal, A. Prakash, R. Tripathi, and K. Naik, "Regional super cluster based optimum channel selection for CR-VANET," *IEEE Transactions on Cognitive Communications and Networking*, 6(2) (2020): 607–617. doi: 10.1109/TCCN.2019.2960683
9. Raghavendra Pal, Arun Prakash, and Rajeev Tripathi, "Triggered CCHI multichannel MAC protocol for vehicular ad hoc networks," *Vehicular Communications*, Elsevier, 12 (2018): 14–22. doi: 10.1016/j.vehcom.2018.01.007
10. Raghavendra Pal, Arun Prakash, Rajeev Tripathi, and Dhananjay Singh, "Analytical model for clustered vehicular ad hoc network analysis," *ICT Express*, Elsevier, 4(3) (2018): 160–164. doi: 10.1016/j.icte.2018.01.001
11. Pant Varun Prakash, Saumya Tripathi, Raghavendra Pal, and Arun Prakash, "A slotted multichannel MAC protocol for fair resource allocation in VANET," *IJMCMC*, IGI global, 9(3) (2018): 45–59. doi: 10.4018/IJMCMC.2018070103
12. P. Singh, R. Pal, and N. Gupta, "Clustering based single-hop and multi-hop message dissemination evaluation under varying data rate in vehicular ad-hoc network," in Choudhary, R., Mandal, J., Auluck, N., and Nagarajaram, H. (eds) *Advanced Computing and Communication Technologies. Advances in Intelligent Systems and Computing*, vol 452. Springer, Singapore, 2016.

13. U. Prakash, R. Pal, and N. Gupta, "Performance evaluation of IEEE 802.11p by varying data rate and node density in vehicular ad hoc network," *2015 IEEE Students Conference on Engineering and Systems (SCES)*, Allahabad, 2015, pp. 1–5. doi: 10.1109/SCES.2015.7506457
14. B. B. Dubey, N. Chauha, N. Chand, and L. K. Awasthi, "Priority based efficient data scheduling technique for VANETs," *Wireless Networks*, 22(5) (2016): 1641–1657. doi: 10.1007/s11276-015-1051-8.

9

An IOT-Based Accident Detection and Rescue System – A Prototype

M. J. Vidya, K. Veena Divya, and P. M. Rajasree

CONTENTS

ABSTRACT Road accidents are one of the major causes of death in India, and this death rate is increasing every year. Not all accident victims lose their lives on the spot; some victims die due to delayed response times from rescue teams. Hence the main aim of this chapter is to devise a method to minimize the lives lost every year due to road accidents. In current systems, the people who are present at the site of an accident are responsible for informing the hospital about the accident so that an ambulance can be sent to save the victims. But the situation is critical since the time needed to rescue a seriously-injured person is precious.

To improve the current situation, the system developed in this work can be installed in vehicles. It is capable of detecting an accident, acquiring the location of the accident, and informing the control center about the accident and its location. It can also send real-time information about the patient once the ambulance arrives at the accident site, and the victim is transferred to the ambulance. The vibration sensor detects the accident, GPS acquires the location, and this location will be shared through the GSM module. Since prevention is better than cure, this system includes a gas sensor which is used to detect the presence of alcohol (in the case of drunk driving). If alcohol is detected, the vehicle's engine will be turned off. The system includes a manual switch, which is to be pressed by the accident victim if the accident is not severe, and, hence, the victim does not require a rescue team. If the victim does not press the manual switch, the accident location will be sent to the control center. The nearest hospital, with respect to the shared location of the accident, will be contacted and an ambulance can be dispatched to the site of the accident. After reaching the location of the accident, the ambulance can transmit real-time information about the patient to the hospital, like the patient's heart rate. To reduce the delay of ambulances en route to the hospital, a zero-traffic system is implemented by transmitting the RF signal to the traffic signal controller.

9.1 Introduction

India is facing serious road accident problems because of a high motorization growth rate, which is accompanied by urbanization and rapid extension to road networks over the years. Road accidents are influenced by many contributing factors, such as length of the road network, enforcement/adherence to road safety regulations, and human population. Road accidents result in injuries, disabilities, fatalities, and hospitalization with a huge concern both at a national and international level.

Accidents in India, as well as worldwide, are one of the leading causes of death every year. In India, every year around 130,000 human lives are lost and 420,000 people are injured in road accidents [1]. Figure 9.1 shows the national figures of deaths in road accidents from 2011–2018 [1] on national highways, state highways, and others in India.

The overall number of road accidents has increased by 1.4% from 2017 to 2018. The accident severity (number of people killed per hundred accidents) has been increasing year after year. As per the recent survey done by Vehicles and Road Traffic, Statista, 2020, Figure 9.2 shows an increase in the number of fatalities due to road accidents from 2005–2018.

The study of road accident data shows that 150,000 deaths occur due to accidents in the country on Indian roads [2]. This is a real concern to be dealt with as a matter of utmost importance.

9.2 Literature Survey

With India's rapid population growth rate, vehicles have become an outright need for everyday life, which results in multiple road accidents and

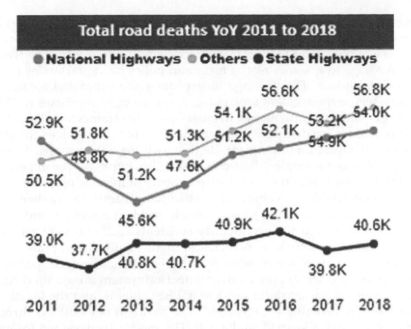

FIGURE 9.1
National figures of deaths in road accidents from 2011–2018 in India.

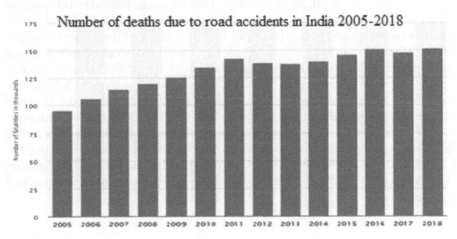

FIGURE 9.2
Number of deaths due to road accidents in India from 2005 to 2018. (Courtesy: Statista Research Department, February 2020.)

a resulting large death toll, which is due to the gap between the time when the rescue team receives the information and the time required by the ambulance to reach the accident spot. The automatic accident detection and human rescue system (AADHRS) [1] proposes a system that identifies vehicle accidents and helps save lives by educating salvage groups. Vibration sensors, global system for mobile (GSM), and global positioning system (GPS) are systems that are utilized at the moment. In the wake of obtaining the location of accident, the framework conveys a short message to a nearby rescue team and police headquarters by means of a GSM module. The message incorporates the estimated location of the area where the accident took place. A rescue team then uses the location to help injured individuals. In most cases, this framework[3] contains various sensors for distinguishing accidents, which minimizes the effort to search the location of accident. Be that as it may, the proposed model incorporates just a single vibration sensor; this bypasses the expense of numerous sensors and the multifaceted nature of interfacing. In this way, it will be affordable for vehicle owners in Bangladesh. In another paper [4], an internet of things (IoT) based vehicle accident discovery and rescue framework is created so as to identify accidents and send the location of the mishap to the vehicle owner, closest medical clinic, and police headquarters by means of a web administration system. The correspondence between the web server and accident detection system is built up through GSM or general packet radio services (GPRS) shields, and the location is pinpointed by utilizing a GPS shield. The accident is identified through vibration sensors, a keypad, and a bell. The module is designed for gathering accident location information continuously using a web application, android versatile application or SMS. This module provides the precise location of the area where the accident occurred and sends a warning notification to the closest police headquarters and clinic. An accident recognition system [5] was built which utilizes an on-board accelerometer sensor to recognize accidents and produce crisis caution and alert the emergency contact. Rescue efforts are facilitated by the continuous tracking[6] of the locations of both the casualty and emergency contact. Literature reviews suggest the use of different systems to rescue the victim of an accident as soon as possible. But the proposed system in this work suggests a better solution of saving the victim by alerting the control room, acquiring the biomedical parameters of the victim in the ambulance, and maintaining zero traffic for the ambulance to reach the nearest hospital in the shortest time possible.

9.3 Research Gap

Existing solutions require an accident victim's action after an accident occurs. These mobile solutions expect the injured person to launch an application and request help manually, which would not be possible if the person was in a critical condition. The situation is even worse if the person becomes unconscious.

Controlling road traffic signals is one of the most important issues in a rescue system since each and every second counts when saving a life. This is something which has never been addressed before.

9.4 Motivation

Road accidents are one of the major causes of death in India, and this death rate is increasing every year. Not all motor vehicle crash fatalities occur at the moment of impact; some fatalities are caused due to delayed response time from emergency teams. Some rescue teams face difficulties in reaching the injured people due to delayed alerts and inadequate information on the exact accident location. In the case of brutal accidents, the victims are not able to call for assistance; and on secondary roads, vehicles may be difficult to find by a rescue team. Hence there is a need for an automated system which would solve these problems. The death rate due to road accidents in India from 2011–2018 is shown in Table 9.1.

It is clear that, owing to the high total road deaths, there is a need to develop an automated system that will help save many lives.

TABLE 9.1

Share of Road Categories in Road Accidents: Number of People Killed and Injured from 2008 to 2018 [1]

	National highways			State highways		
Year	Road accidents	Persons killed	Persons injured	Road accidents	Persons killed	Persons injured
2008	28.5	35.6	28.6	25.6	28.4	27.5
2009	29.3	36.0	29.6	23.8	27.1	25.5
2010	30.0	36.1	31.3	24.5	27.3	26.0
2011	30.1	37.1	30.5	24.6	27.4	26.1
2012	29.1	35.3	30.1	24.2	27.3	25.9
2013	28.1	33.2	28.9	25.6	29.6	27.6
2014	28.2	34.1	29.9	25.2	29.1	26.8
2015	28.4	35.0	29.1	24.0	28.0	26.3
2016	29.6	34.5	29.6	25.3	27.9	25.8
2017	30.4	35.9	303	25.0	26.9	25.4
2018	30.1	35.7	30.0	25.2	26.8	25.9

9.5 Aim

To develop a prototype for an automatic accident detection system to track the location of accidents and alert emergency services to save victims.

9.6 Key Contributions

1. Development of a new, smart, working prototype solution to help society in reducing the death rate resulting from vehicular accidents.
2. Assurance that passenger intervention is not required after the accident.
3. Collection of geographical data regarding the accident location.
4. Automatic transmission of bio-signal information required by the rescue teams to the hospital.
5. Turning off the engine if alcohol is detected in the vehicle.
6. Maintaining zero traffic for ambulances along the way to the hospital.

9.7 Methodology

Accident detection is a task which is being extensively used these days for the safety of passengers of a moving vehicle. One of the places where this method is used is in the employment of airbags in the event of a major accident. It was in the year 1984 that the Federal Government of the US made it compulsory for all vehicles manufactured after 1989 to be fixed with an airbag system or a supplementary restraint system (SRS). Also, the likelihood of detecting accidents with a range of sensors led to the development of various systems that would advance the chances of survival of passengers involved in an accident.

Telemedicine is one of the sophisticated technologies of the 21st century. It can be used to provide supplementary medical services and has accordingly been used in emergency situations, for personal healthcare, in mobile hospitals, by rapidly alerting doctors to a patient's rehabilitation, disease, etc.

By exploiting the advances of wireless multimedia communications, such as various current utilities, convenience, network availability and reach, and high data transmission rates, injured victims are able to be transported to emergency healthcare centers as quickly as possible. This allows the physician in the clinic to access information about the patient's condition in advance and arrange for the necessary emergency medical resources, which could decrease the actual time required for treating the victim.

Block diagram of the prototype

A block diagram of the automatic detection and rescue system is shown in Figure 9.3. The entire system can be divided into four main modules, namely the vehicle module (VM), ambulance module (AM), zero-traffic module (ZTM), and the control center (CC). The VM has a gas sensor (MQ5) to detect the presence of alcohol, a vibration sensor (SEN263) to detect an accident, GPS to acquire the accident location, and GSM through which the information will be shared. Wi-Fi is used to transmit the information to the control center. The micro-controller (ARM7 LPC2148) is used here due to its many advantages over other controllers.

When a vehicle is in an accident, the vibration sensor detects the impact of the accident, and the GPS acquires the location, which is shared through the GSM module. The nearest hospital can be contacted, and an ambulance can be sent to the exact spot. After reaching the scene of the accident, the ambulance can transmit real-time health information about the patient, such as the patient's heart rate. Prevention is better than cure; hence in this system, a gas sensor (MQ5) is used to detect the presence of alcohol, and if detected, the engine of the vehicle will be turned off. The system would include a manual switch, which can be activated in the case of a minor accident where the victim does not require any help from a rescue team. To reduce delays in reaching the hospital, zero traffic will be maintained along the way to the hospital.

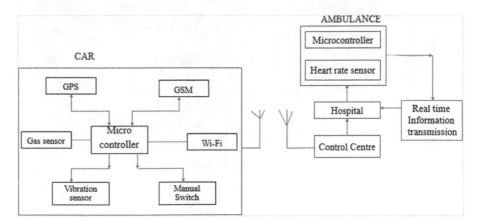

FIGURE 9.3
Block diagram of automatic accident detection and human rescue system.

The ambulance could be equipped with ambulatory devices such as pulse-oximetry [7–8] and other patient monitoring devices to alert the hospital before the patient arrives.

9.8 Hardware Implementation of the Proposed System

As discussed in Section 9.7, the hardware unit of the IOT-based automatic accident detection and rescue system can be divided into the following modules: Vehicle module, ambulance module, and zero-traffic module. The interfacing diagram is shown in Figure 9.4.

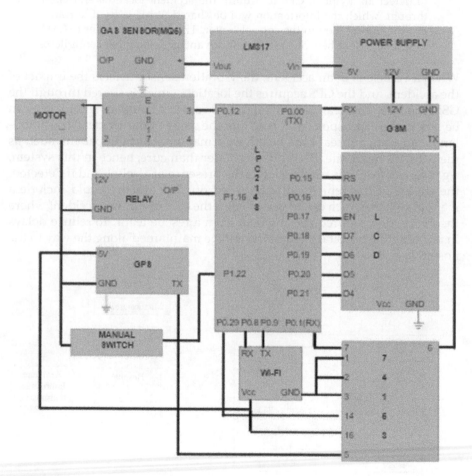

FIGURE 9.4
Interfacing module of the vehicle module.

The vehicle module developed in this work needs to be installed inside the vehicle, close to the driver's seat. The vibration sensor (SEN263) buffers a piezoelectric transducer and generates voltages. The sensor recognizes the vibration during the accident, the GPS acquires the accident's location and uses GSM through which the accident information will be shared. The baud rate of the GSM module is adjusted at 9600 using AT commands. Drunk driving is also one of the major causes of road accidents. To address this issue, a gas sensor (MQ5) is used to detect the presence of alcohol. The gas sensor has a Ni-Cr heating coil, and it has a sensitivity of 100 parts per million (PPM) for alcohol. When it detects alcohol, the molecules undergo ionization and the voltage increases. Once the voltage increases above the reference value, the output of the comparator becomes high, which is considered as the output of the gas sensor. To turn the engine on or off, the project makes use of a switch which will be in a normally closed (NC) position initially and then moves to a normally open (NO) position when the gas sensor detects alcohol and turns off the engine.

The ambulance module makes use of a microcontroller (LPC2148) and motor controller (L239D) which is used to drive DC motors of the ambulance wheels, a heart rate sensor [9,10], an RF transmitter (R8C27), two manual switches to control traffic signals, and an LCD to display messages. The motor driver (L293D) consists of two enable pins (IN1 and IN2 associated with EN1 and IN3, and IN4 associated with EN2). Either of the input pins should be set low to run the motor; otherwise the motor will not run.

The RF transmitter is used to control the traffic signal, and it requires an input voltage of +5 V. Two switches are transmitted with this transmitter; one is connected to pin0, and the other is connected to pin1 of the transmitter. Upon pressing these switches, the RF receiver mounted near the traffic signal receives the signals and controls the traffic accordingly. The LCD is used to display the status and the heart rate of the victim. The measured heart rate and other real-time data of the patient can be directly sent to the hospital in order to monitor the patient's condition well before he/she reaches the hospital, allowing hospital staff to make the necessary arrangements.

For testing the system, the components of the ambulance module were mounted on a metal board with wheels so that it was able to move. It was the closest to the real vehicle for a working demonstration.

The zero traffic implementation module consists of the RF receiver, microcontroller (8051) and the traffic signal lights (r – red, a – amber, and g – green). The microcontroller communicates between the RF receiver and the traffic signal at a frequency of 424 MHz. The transmitter is placed in an ambulance and the receiver near the signal side.

9.9 Software Implementation of the Proposed System

Keil uVision4 software has been used for programming the ARM7-based microcontroller LPC2148. Embedded C programming language and Flash Magic have been used to dump the code into the microcontroller.

9.10 Results

This section discusses the results obtained while testing the prototype for the automatic accident detection and human rescue system. Whenever an accident occurs, the location of the accident site will be shared with the control center and it will also be sent to the registered mobile number. The person managing the control center needs to inform the nearest hospital about the accident so that the ambulance can receive the accident site. In the case of a small accident (where the people in the vehicle are not seriously injured), a manual switch is pressed by the driver and the accident location is not shared. When the system detects the presence of alcohol, the engine is turned off and thus helps to prevent drunk driving, which is one of the major causes of road accidents in India.

Screenshots of the messages received by registered mobile numbers and the control center are shown in Figure 9.5.

The system was tested in various places, and the latitude and longitude of different accident locations were sent to nearby hospitals and to the control center.

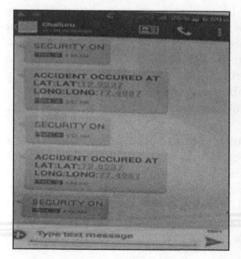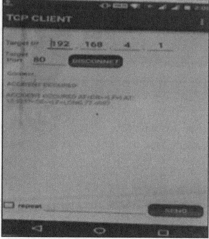

FIGURE 9.5

Message from accident location sent from registered mobile number to the control center.

Figure 9.6 shows the results when an accident occurs, but the accident victim presses the manual switch because he or she is not injured and, hence, does not need help from a rescue team. Figure 9.7 shows the message obtained when alcohol is detected in the system. The engine of the vehicle will be turned off automatically. The system also implements zero traffic in order to reduce delays in reaching the hospital, since each and every second counts when saving a life. If an approaching traffic signal is red, the ambulance driver can press a manual switch provided, which will change the signal to amber and then to green.

9.11 Conclusion

An automatic accident detection and human rescue system has been developed and tested successfully. The entire system can be divided into four

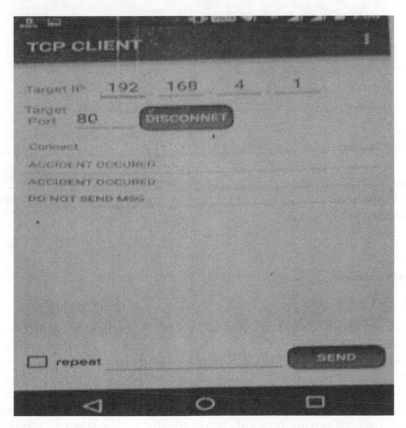

FIGURE 9.6
Rescue messages not sent due to less severe accidents.

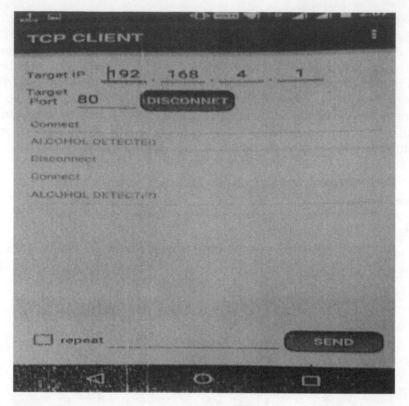

FIGURE 9.7
Message obtained when alcohol is detected.

major modules: Vehicle module, ambulance module, control center, and zero-traffic implementation module. Each module works successfully to save the victim at the right time.

9.12 Future Scope

As already discussed, around 130,000 lives are being lost every year to road accidents in India. This death rate can be reduced to a great extent if the developed system is implemented. In future, real-time cameras can be incorporated in the ambulance for live streaming of videos to the hospital so that necessary steps can be taken by doctors before the accident victim reaches the hospital. A server can be developed to store information of all hospitals and search nearby hospitals automatically with respect to the shared location.

References

1. "Summary of accidents and deaths trend on Highways- National Figures," Ministry of Road Transport & Highways, Government of India, 2018.
2. "Number of deaths due to road accidents across India from 2005 to 2018," Vehicles and Road Traffic, Statista, February 2020.
3. Taj, F. W., Masum, A. K. M., Taslim Reza, S. M., Kalim Amzad Chy, M., and Mahbub, I., "Automatic accident detection and human rescue system: Assistance through communication technologies," *2018 International Conference on Innovations in Science, Engineering Technology (ICISET)*, Chittagong, Bangladesh, 2018, pp. 496–500.
4. Nalini, C. and Swapna Raaga, N., "IoT based vehicle accident detection & rescue information system," *Eurasian Journal of Analytical Chemistry* 13(3) (2018): em2018159. doi: 10.29333/ejac/2018159
5. Khan, A., Bibi, F., Dilshad, M., Ahmed, S., and Ullah, Z., "Accident detection and smart rescue system using Android smartphone with real-time location tracking," *International Journal of Advanced Computer Science and Applications* 9(6) (2018): 341–355.
6. "LPC2141/42/44/46/48- Datasheet," Keil, 2005. https://www.keil.com/dd/d ocs/datashts/philips/lpc2141_42_44_46_48.pdf
7. Vidya, M. J, Padmaja, K. V., and Chaithra, G., "An indigenous PPG acquisition system using monolithic photo-detector OPT101", *IEEE, International Conference on Signal Processing, Communication, Power and Embedded System (SCOPES-2016), Centurion University from 3rd to 5th October, 2016 in Paralakhemundi*, Odisha. doi: 10.1109/SCOPES.2016.7955681
8. Vidya, M. J., and Divya, A. N., "Implementation of cost-effective wrist-based BP monitor," The National Conference on *"Information and Communication Technologies"*, 2016, Department of TCE, R.V. College of Engineering, Bengaluru.
9. Vidya, M. J., and Padmaja, K. V., "Appending photoplethysmograph as a security key for encryption of medical images using watermarking," In: *Progress in Advanced Computing and Intelligent Engineering*, Advances in Intelligent Systems and Computing, Vol. 713, Springer, Singapore, 2019.
10. Vidya, M. J., and Padmaja, K. V., "Enhancing security of electronic patient record using watermarking technique," Science Direct, *Materials Today Journal* 5(4) (2018): 10660–10664. Elsevier. doi: 10.1016/j.matpr.2017.12.341

10

Intelligent Mobility for Minimizing the Impact of Traffic Incidents on Transportation Networks

Banishree Ghosh

CONTENTS

ABSTRACT   Due to major development in urbanization and the expanding population, traffic congestion has become a critical problem in metropolitan cities. Traffic congestion happens whenever the demand exceeds the maximum capacity of a road. Therefore, to make optimum use of the road network capacity, intelligent transport systems (ITS) are often employed to cope with this situation efficiently. Using advanced sensor technologies, real-time traffic information is collected from large transportation networks and utilized in various applications, such as route guidance, congestion avoidance, traffic control and management, etc. Apart from regular instances of peak-hour congestion, the unexpected occurrence of non-recurrent traffic events such as accidents, vehicle breakdowns, and road crashes causes about 25% of traffic congestion on arterial roads and an even higher proportion for urban highways and expressways. Every year, 1.35 million people die on average as a result of road traffic crashes. In addition, 20–50 million people suffer from non-fatal injuries, sometimes leading to short-term or long-term disability. Moreover, non-recurring incidents lead to significant economic losses because of their unpredictable nature. Road accidents can cost up to 3% of a country's gross domestic product. Therefore, anticipating such events in advance can be highly useful in mitigating the resultant congestion and therefore benefiting the national economy as a whole. However, since these types of incidents are non-recurrent and unplanned, the probability of occurrence of these incidents is hard to forecast. Therefore, ITSs are more concerned with minimizing the severity of congestion after the incidents have already occurred. The two integral systems of ITS are traffic information management systems (TIMS) and dynamic routing guidance systems (DRGS). Both of these systems play a vital role because TIMS is responsible for real-time data acquisition of traffic parameters, like speed, the number of vehicles passing by, weather conditions, and DRGS help commuters to dynamically choose a route by providing information on network traffic and other possible routes to be taken. The two techniques used for traffic prediction are either simulation based or data driven. In a simulation-based approach, traffic prediction models which predict the future state are designed based on some theoretical models. This approach needs some expertise to build network traffic simulation. On the other hand, data-driven models can be built for prediction with the usage of historical or real-time data sets.

Apart from the predictive solutions, there are several other new technologies to assist drivers in the occurrence of an incident. For example, traffic management authorities have installed the new age variable message signs (VMS displays) with modern technologies (such as graphics and more colors) on the roads of cities. These LED road traffic signs notify drivers about any kind of disruption in traffic, such as accidents, obstacles, and roadworks, and therefore help in rerouting vehicles. Nowadays, VMS systems form an integral part of DRGS. Therefore, traffic management authorities from several smart cities have been investing a significant amount of resources in installing VMS displays in different places. Thus, ITS are often being employed in different aspects of transportation to provide better mobility solutions for commuters and drivers. Moreover, advancements in sensor technologies have helped ITS to improve the efficiency of existing transportation infrastructure. In this chapter, we address the applications of ITS for incident management and congestion avoidance in real time.

10.1 Introduction

Non-recurrent traffic incidents such as accidents, vehicle breakdowns, crashes, obstacles, etc. lead to severe congestion in metropolitan cities and therefore have a major impact on roadway traffic. These incidents affect the daily lives of commuters either directly or indirectly. Moreover, due to the ever-growing population, the need for intelligent and practical traffic management systems arises that would provide better traffic congestion control in the event of a traffic incident [1]. Of late, intelligent transportation systems are being employed to study traffic incidents systematically and identify their root causes [2]. Apart from adverse geographical or environmental conditions, human errors are also responsible for a large percentage of traffic incidents across the world [3][4].

Since these incidents are non-recurrent and unplanned, they cannot always be predicted in advance and, as such, there are many false-negative cases. Therefore, recent studies focus on building predictive models for minimizing the impact and severity of the incidents after they have already taken place. Moreover, smart technologies, such as variable message signs (VMS) or variable speed limit (VSL), are also in operation to assist drivers in major cities with the aim of improving safety [5][6]. Thus, intelligent mobility has come into play for minimizing the overall traffic disruption caused by incidents and thereby facilitating better travel for commuters and transport management.

10.2 Different Types of Incidents

Based on the geographical location of cities or countries, several types of incidents can be found in the literature. For example, Araghi *et al.* [7] identified ten types of incidents in the Greater London Area: Broken-down vehicle, broken-down lorry, accident, flood, fuel spillage, gas leak, fire, police incident, collapsed manhole, and traffic light failure. Each of these incident types may have several sub-categories. For example, accidents can be of different kinds: Rear-end collision, side-impact collision, head-on collision, rollover, single-car accident, multiple vehicle pile-up, etc. [4]. The annual number of deaths per 100,000 people due to road accidents in 2017 is shown in Figure 10.1 for different countries all over the world. In general, the types of incidents vary with the traffic conditions (i.e., speed limit, traffic flow), geographical attributes (i.e., the width of the roads, capacity), as well as weather (i.e., rainfall, fog, snow). Moreover, the types of incidents recorded from highways or freeways can be significantly different from the incident types that are recorded from arterial roads. Considering all these variations, we can broadly classify the traffic incidents into six categories: Vehicle breakdown (disabled vehicles), accident (crashes and collisions), obstacle on the road (stationary and abandoned vehicles), roadblock or diversion, car on fire, and miscellaneous (which includes other types of incidents such as fuel spillage, traffic light failure, etc.). These types of incidents are mostly common for all cities, as mentioned in past studies [1][9].

Death rate from road accidents, 2017
The annual number of deaths from road accidents per 100,000 people.
Deaths include those from drivers and passengers, motorcyclists, cyclists and pedestrians.

No data   0   5   10   15   20   25   30   40   50   60   >70

Source: IHME, Global Burden of Disease (GBD)
Note: To allow comparisons between countries and over time this metric is age-standardized.

FIGURE 10.1
Annual death rate from road accidents in 2017 [8].

10.3 Causes of Incidents

Traffic incidents can be caused due to several reasons. Although adverse environmental factors or poor vehicle conditions may lead to traffic incidents, humans can also be responsible for a large percentage of traffic incidents [3]. Below are the most common causes of traffic incidents:

1. *Drivers*: There are many laws and rules concerning driving in every country since in many cases drivers are solely responsible for road accidents and crashes. Distracted driving brings about 25% of the incidents all over the world, which includes texting, eating, talking, or doing other tasks while driving. Apart from these causes, drunk driving and sleep deprivation are even more dangerous and pose a great threat to all road users. In addition, reckless or rash driving, speeding, and violation of traffic rules and traffic lights can all lead to mishaps on the roads.

2. *Defects in the vehicle*: Vehicle breakdowns are mostly caused by flaws in the vehicles, such as brake failure, failure of steering system or lighting system, tire burst, etc. Therefore, the owners of the cars should take their cars for regular maintenance to avoid such incidents on roads.

3. *Poor environmental conditions*: Rainfall, snowfall, fog, etc. cause hindrances in smooth driving, and therefore may lead to traffic incidents. Due to rainfall and snowfall, the roads become slippery. As such, vehicles tend to skid more due to lower friction with the roads. Moreover, poor visibility due to heavy precipitation or fog, mist, smoke, etc. makes driving unsafe [10].

4. *Road obstructions*: Sometimes, trees or branches can partially or fully block roads. Moreover, there can be broken down vehicles on the roads which may require road users to brake abruptly. Apart from that, cones placed on the roads due to roadworks, animals crossing the roads, etc. can also lead to accidents. Although there are many other potential scenarios causing traffic hazards, these are the major causes of traffic incidents in different cities across the world.

10.4 What Is Intelligent Mobility?

Intelligent mobility is a buzzword nowadays, which encompasses the technological solutions for communication, control, and information processing in transportation [11]. With the growing number of vehicles on the road, traffic safety has become of utmost concern. Therefore, vehicles are now fitted with new technologies which can help avoid crashes and minimize the after-effects and therefore ensure road safety. Moreover, traffic management authorities are also being proactive in maintaining the smooth operation of traffic. They are adopting new technologies to inform road users about roadworks or other obstructions on the roads [12]. Nevertheless, traffic incidents are inevitable mostly because of human error or faulty vehicle parts [4]. To circumvent such situations, real-time traffic prediction models have been developed, which can forecast the incident duration and spread of congestion in advance so that road users can avoid getting stuck in traffic jams [13]. Thus, the overall congestion caused by the incidents can also be minimized.

10.5 Incident Duration Prediction

To optimize and improve the incident management strategies, the duration of the incidents is an important parameter to be considered. Predicting an incident's duration in real-time can be highly useful for driver assistance systems. Based on the predictions, they can provide advisories to drivers to avoid the incident-affected location for the remaining duration of a trip. Therefore, incident duration prediction has always been an active research area in transportation.

The total incident duration comprises four major stages: Reporting time (the time taken to detect, verify, and report the incident after its occurrence), response time (the time taken by the response team to arrive at the spot after reporting of the incident), clearance time (the time required by the team to clear the affected area), recovery time (the time taken by the traffic condition to restore back to normal) [14]. However, the standards and definitions vary from one country to another. Some studies defined incident duration in terms of the first three stages since the incidents get cleared by the end of the clearance time [15]. A few other studies considered the reporting time (i.e., when the incidents had been reported) as the starting time of the incidents. As per these studies, the span of the incidents includes the response time and the clearance time, i.e., the second and third stages [16]. Nevertheless, most of the studies took all four stages into consideration since the impact of the incidents exists in the network from the beginning of the reporting time (first stage) to the end of the recovery time (fourth stage) [17].

The existing studies relating to incident duration prediction are enlisted in Table 10.1.

TABLE 10.1

Summary of Previous Studies on Incident Duration Prediction

Type of model	Methodology used	Literature	Target variable	Error
Hazard-based model	Log-logistic model	Chung *et al.* [18]	Accident duration	MAPE value 47%
	Gamma, Weibull and log-logistic models	Jiang *et al.* [19]	Incident duration	40% of incidents having <10 min error
	Mixture model and AFT model	Ruimin *et al.* [20]	Incident duration	MAPE 45% for >15 min
Traditional machine learning	Hybrid tree-based quantile regression	Qing *et al.* [14]	Incident duration	MAPE value 49.1%
	ANN, SVM/RVM, KNN	Valenti *et al.* [21]	Incident duration	MAPE in the range of 34–44%
	Text analysis and ANN	Pereira *et al.* [22]	Incident duration	MAPE varies over time from 100% to 40%
	SVR	Wu *et al.* [23]	Incident duration	Total accuracy 70%
	Bagging and boosting, SVR, and MLP	Banishree *et al.* [17]	Incident duration	MAPE varies over time from 61% to 27.85%
Combined model	Combined M5P tree and HDBM	Lin *et al.* [24]	Accident duration	39.1% and 33.15% for two different data sets
	Combination of ANN models	Lopes *et al.* [25]	Clearance time	Mean error: 10–20 min

In earlier years, most of the studies applied simulation-based theoretical modeling for traffic prediction. For example, Golob *et al.* [26] developed log-normal models of incident duration, whereas Hojati *et al.* [9] built prediction models with Weibull and log-logistic distributions, as shown in Figure 10.2. Later, a few studies found that incident duration can be better approximated if separate statistical models are fitted to different stages of the incidents. For example, Ruimin *et al.* applied separate statistical models for each stage, such as dispatch time, clearance time, etc. [27], and subsequently developed a mixture model combining generalized gamma, Weibull, and log-logistic distribution for different incident clearance methods [20]. However, all these studies developed static models which did not incorporate real-time traffic information.

In recent years, various data-driven modeling algorithms, such as support vector regression (SVR) [23], artificial neural networks (ANN) [13], and bagging and boosting [28], have been applied for predicting the duration of traffic incidents. Moreover, the adaptive network-based fuzzy inference system (ANFIS) has proved to be efficient because it combines the efficiency of the fuzzy model in handling uncertainty with the prediction accuracy of the ANN method using back-propagation [29].

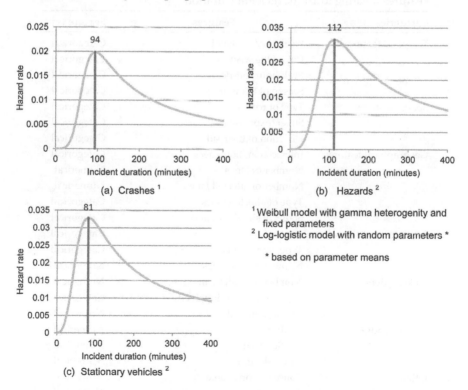

(a) Crashes [1]

(b) Hazards [2]

(c) Stationary vehicles [2]

[1] Weibull model with gamma heterogenity and fixed parameters
[2] Log-logistic model with random parameters *

* based on parameter means

FIGURE 10.2
Hazard functions for different incident types [9].

However, an important consideration in building data-driven models is the comprehensiveness of the data set. The incident duration depends on several external factors, such as spatiotemporal features, geographical features, incident information, stakeholders, traffic conditions, etc. Examples of each category of features are mentioned in Table 10.2, although the list is not exhaustive [30].

Among these features, real-time traffic information is one of the most important features since the forecast should be updated every time the traffic condition changes. Therefore, most of the recent studies carried out have analyzed the impact of incidents by incorporating real-time traffic data [13] [14]. In particular, it is essential to consider both important traffic variables, i.e., speed and flow, in the analysis which can be explained by the fundamental diagram of speed-flow, as shown in Figure 10.3.

The relationship between the variation of speed and flow can be described as follows. The flow can be low because either there is not much traffic on the road or the vehicles cannot move because of congestion. Now, when there is

TABLE 10.2

Features Relating to Traffic Incident Duration

Categories	Features	Feature type
Temporal features	Weekday/weekend	Categorical
	Day of the week	Categorical
	Peak-hour/off-peak	Categorical
	Season of the year	Categorical
Spatial features	Type of road	Categorical
	Street/expressway	Categorical
	Direction of the road	Categorical
Geographical features	Intersection/bottleneck	Categorical
	Number of lanes	Numerical
	Number of affected lanes	Numerical
	Type of affected lanes	Categorical
	Status of shoulder lane	Categorical
Incident characteristics	Incident severity	Categorical
	Type of incidents	Categorical
	Number of casualties	Numerical
Stakeholders	Number of vehicles involved	Numerical
	Types of vehicles involved	Categorical
	Fatality/injured	Numerical
Traffic condition	Traffic flow	Numerical
	Traffic speed	Numerical
	Queue-length	Numerical
Other	Police response time	Numerical
	Arrival of emergency response service	Numerical
	Type of reporting	Categorical

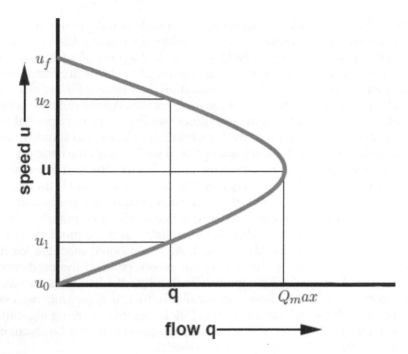

FIGURE 10.3
Speed-flow diagram [31].

not much traffic, the vehicles can move at almost free-flowing speed. By contrast, if the flow is low due to congestion, the average speed of the vehicles will be low as well. Therefore, from the fundamental speed-flow relationship [31], we can deduce that if both the speed and flow are lower than usual, the road is clearly congested. Moreover, rules based on a single traffic variable do not generalize well to diverse city-scale networks. For example, the speed limit of arterial roads is usually lower than that of expressways. Therefore, the speed will drop low during incidents on arterial roads compared to expressways. Hence, the change in flow is a better indicator of congestion for the incidents on arterial roads. Thus, considering both speed and flow helps to avoid false alarms raised due to sensor noise and also allows us to design more robust decision rules for detecting congestion. The spatial and temporal resolution of traffic data (the interval of the recorded data) is also important to understand the subtle changes in the impact of incidents accurately. For example, in Banishree *et al.*'s work, the spatial resolution of traffic data is 100 m and the temporal resolution is five minutes [17], which can indicate the start and end of congestion with high precision.

Despite considering all potentially significant features in the model, the possibility of partial information or ever-changing ground conditions makes the task of forecasting particularly challenging. Therefore, the predictive models need to be adaptive to flexibly incorporate features as incident and traffic data gradually become available. Moreover, the model should provide reasonable forecasts even when a limited amount of information is available, which gets more refined with elapsing time [17]. A block diagram of such a dynamic adaptive prediction model is shown in Figure 10.4. The entire feature-set is divided into four subsets of features based on their order of availability. The first set (basic set) comprises the basic features which are available immediately after the incident happens. The second set (basic set and optional set 1) contains the basic features and the total number of lanes and shoulder lanes. Similarly, the third set (basic set and optional set 2) covers the basic set and a few other additional features about the lanes. Finally, the fourth set (all features) includes all the features together. The regression model is trained with each subset individually so that the model can perform the prediction when any of the subset is available. Therefore, when a new data-point comes, the model at first checks which feature set is available, and the prediction can be performed by the corresponding model. Any traditional machine learning method can be applied in this model. Moreover, prior to the prediction, a clustering method can also be applied for dividing the incidents into different clusters based on their latent similarities. In that case, separate regression models are built for individual clusters [32]. Different clustering algorithms, such as k-means clustering [33], affinity propagation [34], and Gaussian mixture model [35], can be employed in this model.

While the model performs predictions, commuters and other users need to have confidence in the predictions so that they can rely on them. It is quite obvious that the reliability of the model will increase with time. In this

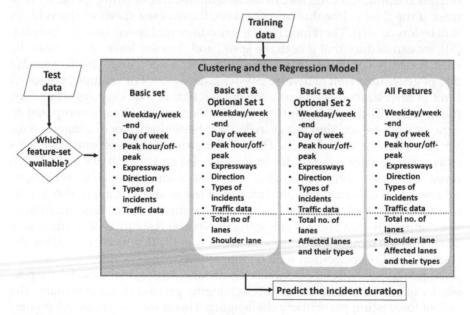

FIGURE 10.4
Block diagram of a dynamic adaptive predictive model.

context, Bayesian methods can be applied for determining the confidence intervals associated with the predictions. In the past, Van *et al.* [36] built a Bayesian neural network model for the prediction of travel time. However, neural network algorithms may sometimes converge on local minima instead of global minima. To solve this issue, other methods, such as naive Bayesian classifier [37] and Bayesian SVR [38][39], have been employed in other studies.

The example in Figure 10.5 demonstrates how a dynamic adaptive incident duration prediction model works in real life. At first, the traffic is normal at around 17:00 on a highway. At 17:05, a vehicle breakdown occurs and blocks the road. Therefore, traffic starts to slow down. The model detects the change in the traffic data and speculates that an incident has happened. Although the incident reports are not available yet (since it takes some time to report the incidents to the management authority), the regression model performs the first prediction with the information it obtains from the traffic data. Meanwhile, the incident is reported at approximately 17:15 (ten minutes after the incident happened). The essential features are disseminated after the incident has been reported, and therefore the database can be updated (i.e., the location, etc.) according to the report. Consequently, the model performs the prediction again at 17:20. Further, the details about the closure of the lanes are reported at 17:30. Therefore, as all features are available by this time, the prediction gets more accurate. In the meantime, the response team reaches the location and clears the damaged vehicle from the road at 17:40. However, traffic usually takes some time to go back to normal after the incident has been cleared. At 17:50, the traffic conditions are recovered although a queue of cars still exists. Finally, at 18:00, the model finds the traffic data to be back to normal again and hence, stops the prediction [17].

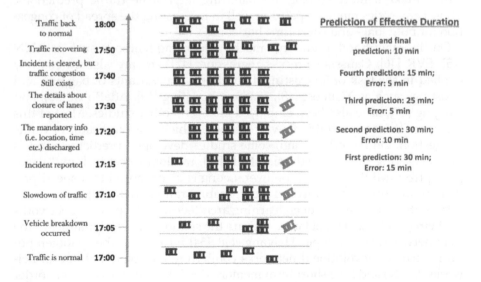

FIGURE 10.5
A real-life scenario demonstrating how the incident duration predictive model works [17].

10.6 Incident Impact Prediction

The modeling and estimation of traffic congestion is another sought-after research area in the field of urban transportation. Nowadays traffic data have become readily available on many government websites since traffic management authorities regularly update the traffic information on their websites collected from a variety of sensors. However, the analysis of traffic congestion is critical because it spreads both spatially and temporally. Moreover, the prediction has to be updated at every instant.

A large number of studies applied the cell transmission model [40] or shock wave model [41] for modeling the congestion propagation in the network. However, none of these studies dealt with real-time traffic data. Hence, the performance of these models may degrade when used in real-time applications. Moreover, the spatial transferability of these models may be compromised as well. Apart from that, several studies have employed time-series approaches, such as the autoregressive moving average (ARMA) and its variants. For example, Alghamdi *et al.* developed the widely used autoregressive integrated moving average (ARIMA) model for forecasting traffic congestion [42], whereas Kumar *et al.* built a seasonal ARIMA (SARIMA) model [43]. Nevertheless, the SARIMA model is more suitable for modeling the recurring congestion because this model can capture the periodicity of the data better. Besides, Kamarianakis *et al.* proposed two multivariate approaches, vector ARMA (VARMA) and space-time ARIMA (STARIMA), for forecasting traffic flow [44]. The VARMA model captures the linear dependence between multiple time series, whereas the STARIMA model requires less parameters, which make it more suitable for handling large-scale traffic predictions. Thus, time-series methods perform well in forecasting the spread of congestion for both long- and short-time horizons.

On the other hand, traditional machine learning methods such as ANN [45], SVR [46], Gaussian process regression [47], etc. are also extensively applied in the task of forecasting. In addition, the variants of bagging and boosting, such as XGBoost [48], gradient boosting [49], AdaBoost [50], and bagging [47], have also been employed in several other studies. Most of this literature focuses on traffic flow prediction, which is an indicator of traffic congestion. On the other hand, some studies developed predictive models for forecasting traffic speed using SVR [51], random forest [52], multilayer perceptron (MLP) [52], etc. Moreover, Leong *et al.* proposed that speed prediction can further be improved incorporating rainfall data [53].

With the emergence of deep learning architectures, deep neural networks are being explored a lot of late to capture the spatiotemporal dynamics of the road network. For example, Haiyang *et al.* [54] introduced the spatiotemporal recurrent convolutional networks combining convolutional neural networks (CNN) and long short-term memory (LSTM) neural networks in order to predict traffic speed for different horizons. Since the LSTM model can

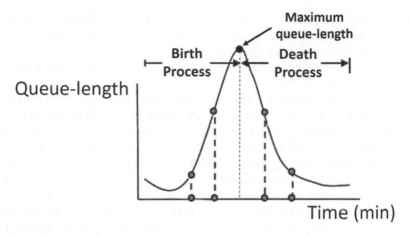

FIGURE 10.6
The birth and death process of congestion [58].

handle the temporal dynamicity of the data efficiently, many studies opted for the LSTM model for traffic-related applications. For example, Min *et al.* [55] introduced a stacked LSTM neural network considering both spatial and temporal correlations to predict traffic speeds, whereas Honglei *et al.* [56] built an LSTM network for predicting the accident risk citywide.

In general, the process of development and dissipation of congestion follows a birth–death mechanism [57], as shown in Figure 10.6. The prediction models have to capture the pattern of the entire birth–death process during the congestion. However, the dynamics of the birth process and death process are different. Therefore, the model needs some time to learn the underlying patterns. For short duration incidents, the model does not get enough time, which leads to high prediction errors for most of the cases. For medium and long duration incidents, the model has sufficient time to learn the birth and death process individually. However, errors may be higher during the switching from birth to death process. The sequence is as follows: The prediction error decreases at first during the birth process until the queue reaches its maximum length, followed by subsequent dissipation while the error is a bit high. Once the model learns about the death process from the previous data, the prediction gets more refined in later instants.

10.7 Other Technologies

For the optimal utilization of road network capacity and efficient traffic management during incidents, several countries have adopted a new technology named VMS. These messages appear on electronic displays in the form of text and graphics to notify drivers about any kind of disruption in traffic, such as accidents, obstacles, roadworks, etc. Moreover, the signs can display the warning and advisories about taking alternative routes or limiting travel speed, and disseminate the information about the duration and location of the incidents [59].

Variable speed limits form part of VMS technology, where it indicates a flexible speed limit of the vehicles on the roads [5]. As opposed to the conventional speed limit signs, which are fixed-valued, the variable speed limits are adjusted in accordance with the present environmental and traffic conditions. Therefore, during traffic incidents, bad weather, or roadworks, the maximum speed limits are reduced compared to that of the optimal condition, and commuters are advised to adhere to the limit accordingly.

It has been proven in several studies that VMS can potentially reduce the number of secondary crashes [12] [60]. Moreover, Dos *et al.* [6] reported that the combination of VMS and variable speed limit techniques can be a potential solution to reduce rear-end collisions. Therefore, apart from predictive solutions, these technologies can also assist drivers in the occurrence of a traffic incident. However, these technologies are still in the nascent stage. The awareness of drivers, which is significantly less than optimal in many cities, compromises the effectiveness of these messages as traffic guidance tools [61]. Therefore, with growing recognition for the efficacy of VMS technology, we hope to see a significant increase in the impact of VMS on overall traffic conditions in the coming years.

10.8 Conclusion

For efficient traffic incident management, it is important to adopt new technologies in urban areas. The prediction of parameters, such as incident duration or congestion length, has proved to be highly effective in this regard. However, the incident management authorities need to ensure that commuters and drivers install those applications in their systems and follow the guidance provided. Moreover, the prediction models should be robust, accurate, and adaptive to the varying ground conditions. Apart from that, of late, the VMS system has been an integral part of the dynamic routing guidance system. Therefore, several smart cities across the world are investing a significant amount of resources in installing VMS displays in different locations.

Last but not the least, prevention is better than cure. Hence, road users need to be careful and responsible enough while driving or walking along roads and, thus, play their part in minimizing the risk of incidents on the roads. Moreover, advanced driver-assistance systems (ADAS) are being employed to assist drivers while driving or parking. The safety features of ADAS are designed to alert drivers or take control of vehicles in the occurrence of an incident [62]. Therefore, ADAS have proved to be efficient in reducing road accidents significantly. There are other technologies associated with ADAS as well, such as lane departure warning systems, automatic lane centering, automated lighting, and incorporating traffic warnings, which can aid in minimizing human errors [63].

References

1. Deo Chimba, Boniphace Kutela, Gary Ogletree, Frank Horne, and Mike Tugwell. Impact of abandoned and disabled vehicles on freeway incident duration. *Journal of Transportation Engineering*, 140(3):04013013, 2013.
2. Junping Zhang, Fei-Yue Wang, Kunfeng Wang, Wei-Hua Lin, Xin Xu, and Cheng Chen. Data-driven intelligent transportation systems: A survey. *IEEE Transactions on Intelligent Transportation Systems*, 12(4):1624–1639, 2011.
3. Eleni Petridou and Maria Moustaki. Human factors in the causation of road traffic crashes. *European Journal of Epidemiology*, 16(9):819–826, 2000.
4. Ali Ahmed Mohammed, Kamarudin Ambak, Ahmed Mancy Mosa, and Deprizon Syamsunur. A review of the traffic accidents and related practices worldwide. *The Open Transportation Journal*, 13(1), 2019.
5. Ellen F Grumert, Andreas Tapani, and Xiaoliang Ma. Characteristics of variable speed limit systems. *European Transport Research Review*, 10(2):1–12, 2018.
6. Cristina Dos Santos. Assessment of the safety benefits of vms and vsl using the ucf driving simulator. University of Central Florida, 2007. https://stars.librar y.ucf.edu/cgi/viewcontent.cgi?article=4146&context=etd
7. Bahar N Araghi, Simon Hu, Rajesh Krishnan, Michael Bell, and Washington Ochieng. A comparative study of k-nn and hazard-based models for incident duration prediction. In 2014 IEEE 17th International Conference on *Intelligent Transportation Systems (ITSC)*, pages 1608–1613. IEEE, 2014.
8. https://en.wikipedia.org/wiki/List_of_countries_by_traffic-related_death_rate.
9. Ahmad Tavassoli Hojati, Luis Ferreira, Simon Washington, Phil Charles, and Ameneh Shobeirinejad. Modelling total duration of traffic incidents including incident detection and recovery time. *Accident Analysis & Prevention*, 71:296–305, 2014.
10. Subasish Das, Bradford K Brimley, Tomás E Lindheimer, and Michelle Zupancich. Association of reduced visibility with crash outcomes. *IATSS Research*, 42(3):143–151, 2018.
11. K Huhtala-Jenks and M Forsblom. Mobility as a service–the new transport paradigm. *Trafik & Veje*: 12–14, 2015.
12. P Borrough. Variable speed limits reduce crashes significantly in the UK. *Urban Transportation Monitor*, 1997.
13. Yuye He, Sebastien Blandin, Laura Wynter, and Barry Trager. Analysis and real-time prediction of local incident impact on transportation networks. In *2014 IEEE International Conference on Data Mining Workshop (ICDMW)*, pages 158–166. IEEE, 2014.

14. Qing He, Yiannis Kamarianakis, Klayut Jintanakul, and Laura Wynter. Incident duration prediction with hybrid tree-based quantile regression. In *Advances in Dynamic Network Modeling in Complex Transportation Systems*, pages 287–305. Springer, 2013.

15. Doohee Nam and Fred Mannering. An exploratory hazard-based analysis of highway incident duration. *Transportation Research Part A: Policy and Practice*, 34(2):85–102, 2000.

16. Xiaolei Ma, Chuan Ding, Sen Luan, Yong Wang, and Yunpeng Wang. Prioritizing influential factors for freeway incident clearance time prediction using the gradient boosting decision trees method. *IEEE Transactions on Intelligent Transportation Systems*, 18(9): 2303–2310, 2017.

17. Banishree Ghosh, Muhammad Tayyab Asif, Justin Dauwels, Ulrich Fastenrath, and Hongliang Guo. Dynamic prediction of the incident duration using adaptive feature set. *IEEE Transactions on Intelligent Transportation Systems*, 20(11):4019–4031, 2018.

18. Younshik Chung. Development of an accident duration prediction model on the korean freeway systems. *Accident Analysis & Prevention*, 42(1):282–289, 2010.

19. Rui Jiang, Ming Qu, Edward Chung, et al. Traffic incident clearance time and arrival time prediction based on hazard models. *Mathematical Problems in Engineering*, special issue: Transportation Modeling and Management, pages 288–294, 2014.

20. Ruimin Li, Francisco C Pereira, and Moshe E Ben-Akiva. Competing risks mixture model for traffic incident duration prediction. *Accident Analysis & Prevention*, 75:192–201, 2015.

21. Gaetano Valenti, Maria Lelli, and Domenico Cucina. A comparative study of models for the incident duration prediction. *European Transport Research Review*, 2(2):103–111, 2010.

22. Francisco C Pereira, Filipe Rodrigues, and Moshe Ben-Akiva. Text analysis in incident duration prediction. *Transportation Research Part C: Emerging Technologies*, 37:177–192, 2013.

23. WW Wu, Shu-yan Chen, and Chang-jiang Zheng. Traffic incident duration prediction based on support vector regression. *ICCTP 2011: Towards Sustainable Transportation Systems*, ASCE Library, pages 2412–2421, 2011.

24. Lei Lin, Qian Wang, and Adel W Sadek. A combined m5p tree and hazard-based duration model for predicting urban freeway traffic accident durations. *Accident Analysis & Prevention*, 91:114–126, 2016.

25. JA Lopes. *Traffic Prediction for Unplanned Events on Highways*. PhD dissertation, Instituto Superior Tecnico (IST), 2012.

26. Thomas F Golob, Wilfred W Recker, and John D Leonard. An analysis of the severity and incident duration of truck-involved freeway accidents. *Accident Analysis & Prevention*, 19(5):375–395, 1987.

27. Ruimin Li. Traffic incident duration analysis and prediction models based on the survival analysis approach. *IET Intelligent Transport Systems*, 9(4):351–358, 2014.

28. Khaled Hamad, Mohamad Ali Khalil, and Abdul Razak Alozi. Predicting freeway incident duration using machine learning. *International Journal of Intelligent Transportation Systems Research*, 18: 367–380, 2020.

29. J-SR Jang. Anfis: adaptive-network-based fuzzy inference system. *IEEE Transactions on Systems, Man and Cybernetics*, 23(3):665–685, 1993.

30. Ruimin Li, Francisco C Pereira, and Moshe E Ben-Akiva. Overview of traffic incident duration analysis and prediction. *European Transport Research Review*, 10(2):22, 2018.

31. Tom V Mathew and KV Krishna Rao. Fundamental relations of traffic flow. https://www.civil.iitb.ac.in/tvm/nptel/512_FundRel/web/web.html#x1-140006, Indian Institute of Technology Bombay, India.

32. Banishree Ghosh, Muhammad Tayyab Asif, Justin Dauwels, Wentong Cai, Hongliang Guo, and Ulrich Fastenrath. Predicting the duration of non-recurring road incidents by cluster-specific models. In *2016 IEEE 19th International Conference on Intelligent Transportation Systems (ITSC)*, pages 1522–1527. IEEE, 2016.

33. Tapas Kanungo, David M Mount, Nathan S Netanyahu, Christine D Piatko, Ruth Silverman, and Angela Y Wu. An efficient k-means clustering algorithm: Analysis and implementation. *IEEE Transactions on Pattern Analysis & Machine Intelligence*, 24(7):881–892, 2002.

34. Brendan J Frey and Delbert Dueck. Clustering by passing messages between data points. *Science*, 315(5814):972–976, 2007.

35. Xiong Liu, Li Pan, and Xiaoliang Sun. Real-time traffic status classification based on gaussian mixture model. In *IEEE International Conference on Data Science in Cyberspace (DSC)*, pages 573–578. IEEE, 2016.

36. CP IJ van Hinsbergen, JWC Van Lint, and HJ Van Zuylen. Bayesian committee of neural networks to predict travel times with confidence intervals. *Transportation Research Part C: Emerging Technologies*, 17(5):498–509, 2009.

37. Stephen Boyles, David Fajardo, and S Travis Waller. A naive bayesian classifier for incident duration prediction. In *86th Annual Meeting of the Transportation Research Board*, Washington, DC, 2007.

38. Jinyoung Ahn, Eunjeong Ko, and Eun Yi Kim. Highway traffic flow prediction using support vector regression and bayesian classifier. In *2016 International Conference on Big Data and Smart Computing (BigComp)*, pages 239–244. IEEE, 2016.

39. Banishree Ghosh, Muhammad Tayyab Asif, and Justin Dauwels. Bayesian prediction of the duration of non-recurring road incidents. In *2016 IEEE Region 10 Conference (TENCON)*, pages 87–90. IEEE, 2016.

40. Jiancheng Long, Ziyou Gao, Xiaomei Zhao, Aiping Lian, and Penina Orenstein. Urban traffic jam simulation based on the cell transmission model. *Networks and Spatial Economics*, 11(1):43–64, 2011.

41. Chun Liu, Shuhang Zhang, Hangbin Wu, and Qiang Fu. A dynamic spatio-temporal analysis model for traffic incident influence prediction on urban road networks. *ISPRS International Journal of Geo-Information*, 6(11):362, 2017.

42. Taghreed Alghamdi, Khalid Elgazzar, Magdi Bayoumi, Taysseer Sharaf, and Sumit Shah. Forecasting traffic congestion using arima modeling. In *2019 15th International Wireless Communications & Mobile Computing Conference (IWCMC)*, pages 1227–1232. IEEE, 2019.

43. S Vasantha Kumar and Lelitha Vanajakshi. Short-term traffic flow prediction using seasonal arima model with limited input data. *European Transport Research Review*, 7(3):21, 2015.

44. Yiannis Kamarianakis and Poulicos Prastacos. Forecasting traffic flow conditions in an urban network: Comparison of multivariate and univariate approaches. *Transportation Research Record*, 1857(1):74–84, 2003.

45. Bharti Sharma, Sachin Kumar, Prayag Tiwari, Pranay Yadav, and Marina I Nezhurina. Ann based short-term traffic flow forecasting in undivided two lane highway. *Journal of Big Data*, 5(1):48, 2018.

46. Manoel Castro-Neto, Young-Seon Jeong, Myong-Kee Jeong, and Lee D Han. Online-svr for short-term traffic flow prediction under typical and atypical traffic conditions. *Expert Systems with Applications*, 36(3):6164–6173, 2009.

47. Banishree Ghosh, Justin Dauwels, and Ulrich Fastenrath. Analysis and prediction of the queue length for non-recurring road incidents. In *2017 IEEE Symposium Series on Computational Intelligence (SSCI)*, pages 1–8. IEEE, 2017.

48. Xuchen Dong, Ting Lei, Shangtai Jin, and Zhongsheng Hou. Short-term traffic flow prediction based on xgboost. In *2018 IEEE 7th Data Driven Control and Learning Systems Conference (DDCLS)*, pages 854–859. IEEE, 2018.

49. Huiwei Xia, Xin Wei, Yun Gao, and Haibing Lv. Traffic prediction based on ensemble machine learning strategies with bagging and lightgbm. In *2019 IEEE International Conference on Communications Workshops (ICC Workshops)*, pages 1–6. IEEE, 2019.

50. Guy Leshem and Yaacov Ritov. Traffic flow prediction using adaboost algorithm with random forests as a weak learner. In *Proceedings of World Academy of Science, Engineering and Technology*, volume 19, pages 193–198. Citeseer, 2007.

51. Muhammad Tayyab Asif, Justin Dauwels, Chong Yang Goh, Ali Oran, Esmail Fathi, Muye Xu, Menoth Mohan Dhanya, Nikola Mitrovic, and Patrick Jaillet. Spatiotemporal patterns in large-scale traffic speed prediction. *IEEE Transactions on Intelligent Transportation Systems*, 15(2):794–804, 2014.

52. Charalampos Bratsas, Kleanthis Koupidis, Josep-Maria Salanova, Konstantinos Giannakopoulos, Aristeidis Kaloudis, and Georgia Aifadopoulou. A comparison of machine learning methods for the prediction of traffic speed in urban places. *Sustainability*, 12(1):142, 2020.

53. Leong Wai Leong, Kelvin Lee, Kumar Swapnil, Xiao Li, Ho Yao Tong Victor, Nikola Mitrovic, Muhammad Tayyab Asif, Justin Dauwels, and Patrick Jaillet. Improving traffic prediction by including rainfall data. In *ITS Asia-Pacific Forum*, volume 14, 2015.

54. Haiyang Yu, Zhihai Wu, Shuqin Wang, Yunpeng Wang, and Xiaolei Ma. Spatiotemporal recurrent convolutional networks for traffic prediction in transportation networks. *Sensors*, 17(7):1501, 2017.

55. Min Chen, Guizhen Yu, Peng Chen, and Yungpeng Wang. *Traffic Congestion Prediction Based on Long-Short Term Memory Neural Network Models. CICTP 2017: Transportation Reform and Change—Equity, Inclusiveness, Sharing, and Innovation.* Reston, VA: American Society of Civil Engineers, 2018. 673–681.

56. Honglei Ren, You Song, Jingwen Wang, Yucheng Hu, and Jinzhi Lei. A deep learning approach to the citywide traffic accident risk prediction. *21st International Conference on Intelligent Transportation Systems (ITSC)* pages 3346–3351. IEEE.

57. Leonard Kleinrock. *Queueing Systems, Volume 2: Journal of Computer Applications*, Volume 66. New York: Wiley, 1976.

58. Akintunde A Alonge and Thomas J Afullo. Rainfall time series synthesis from queue scheduling of rain event fractals over radio links. *Radio Science*, 50(12):1209–1224, 2015.

59. https://en.wikipedia.org/wiki/Variable-message_sign.

60. Chris Lee, Bruce Hellinga, and Frank Saccomanno. Assessing safety benefits of variable speed limits. *Transportation Research Record: Journal of the Transportation Research Board*, 1897:183–190, 2004.

61. Banishree Ghosh, Yuanzheng Zhu, and Ulrich Fastenrath. Effectiveness of vms messages in influencing the motorists' travel behaviour. In *2018 IEEE 22nd International Conference on Intelligent Transportation Systems (ITSC)*, pages 837–842. IEEE, 2018.

62. Umar Zakir Abdul Hamid, Fakhrul Razi Ahmad Zakuan, Khairul Akmal Zulkepli, Muhammad Zulfaqar Azmi, Hairi Zamzuri, Mohd Azizi Abdul Rahman, and Muhammad Aizzat Zakaria. Autonomous emergency braking system with potential field risk assessment for frontal collision mitigation. In *2017 IEEE Conference on Systems, Process and Control (ICSPC)*, pages 71–76. IEEE, 2017.

63. https://en.wikipedia.org/wiki/Advanced_driver-assistance_systems.

Index

Note: Page numbers followed by 't' refer the tables and 'f' refer the figures